Environmental management

Environmental management

New directions for the twenty-first century

Geoff A. Wilson & Raymond L. Bryant
King's College London

First published in 1997 by UCL Press

UCL Press Limited
1 Gunpowder Square
London EC4A 3DE
UK

and

1900 Frost Road, Suite 101
Bristol
Pennsylvania 19007-1598
USA

British Library Cataloguing-in-Publication Data
A CIP catalogue record for this book is available from the British Library.
Library of Congress Cataloging-in-Publication Data are available

ISBNs: 1-85728-462-3 HB
1-85728-463-1 PB

Printed and bound in Great Britain.
By T. J. International Ltd, Padstow, Cornwall.

This book is dedicated to Olivia and Sara

CONTENTS

PART III TOWARDS PREDICTABILITY?

PART IV FUTURE DIRECTIONS IN ENVIRONMENTAL MANAGEMENT

PREFACE

As environmental degradation intensifies around the world, an extensive literature on the causes of, and possible solutions to, such degradation has developed. At the centre of this literature is a recognition that environmental degradation is intimately associated with the intensifying human impact on the environment. Yet, a confusing array of approaches seeking to explain that impact has developed. As such, any new book on the environment must contend with this plethora of perspectives.

What is striking about this growing literature is its lopsidedness. On the one hand, much has been written about the nature and causes of environmental degradation. On the other hand, inadequate attention has been given to explaining how environmental managers around the world relate to the environment and to each other. There is a substantive literature on how environmental managers linked to the state seek to implement official policies and practices, but, and as this book suggests, that literature is weakened often because of a tendency to focus mainly on the state as the only "real" environmental manager.

In contrast to many other books on environmental management, this volume eschews a state-centric approach in favour of a more "inclusive" understanding about environmental management and of environmental managers. It argues that environmental management needs to be understood as a multilayered process in which both state and non-state actors manage the environment. Such a re-evaluation is designed to challenge the traditional view of environmental management. By extension, it suggests the need for a revitalization of the field of environmental management in keeping with a more inclusive approach attuned to the challenges facing contemporary and future environmental management.

The objective of this book, therefore, is to contribute to a new understanding of environmental management. It does not describe environmental problems or specific environmental management practices (which are taken as a given), nor does it proffer technical advice to state environmental managers. This book, therefore, should be of interest to all those interested in a critical appraisal of environmental management. It will also appeal to a broader social science audience interested in the application of social knowledge to the understanding of environmental problems. As Benton & Redclift (1994: 1) have suggested, "the current phase of environmental concern has spawned among social scientists a much wider, more diverse, and more imaginative role for the social sciences in environmental debate". A central contention of this book is that scholars of environmental management need to take on board this important consideration. Indeed, in doing so the suggestion is that the meaning of environmental management itself needs to be reassessed.

The organization of this book reflects this concern. In Part I the analytical framework for a more inclusive understanding of environmental management is proposed. In order to appreciate how environmental managers operate within multi-layered environmental management, it is critical to appreciate how a quest for predictability in the face of social and environmental uncertainty is uppermost in the minds of those managers. As Part II illustrates, the record of the human impact on the environment has been one of intensifying use that has ultimately led to increased uncertainty in environmental management. Part III then analyzes the role of politics, the market, and policies in the efforts of environmental managers to combat uncertainty through the pursuit of predictability in environmental management. Part IV concludes the book by evaluating the possible impact of technological change on the ability of environmental managers to pursue such predictability in the future. It also suggests that a re-evaluation of environmental management as a process also requires fundamental changes in the nature and purpose of environmental management as a field of study.

This book, thus, constitutes an exploration of a more inclusive understanding of environmental management. It aims to sketch the broad outlines of such an alternative, although acknowledging all the while that further research will be needed to elaborate this re-evaluation. The book is intended primarily to focus on questioning existing ways of thinking about environmental management. It is hoped that this will contribute to the development of a new understanding of environmental management, both as a process and as a field of study. The assumption is that environmental management is not in need of yet another study about environmental problem solving, but that it requires a more wide-ranging assessment of what environmental management is – and who environmental managers are – as a basis for the clarification of research in this field.

This book could not have been written without the help of many people. The suggestions and advice of Andrew Warren, Ian Simmons, Guy Robinson and Rob Burton were helpful in clarifying analytical and empirical issues. Participants at the Institute of British Geographers "Environmental Philosophies" session at Strathclyde in January 1996 provided further constructive criticism on some aspects of the analytical framework. Thanks also to James Kneale for useful information on the topic of "virtual geographies".

We owe a particular debt to our colleagues, research assistants and students in the Department of Geography at King's College London. Assistance tracking down hard-to-find sources and empirical data was provided by Grant Ballantine, Fiona Shields and Richard Gauld. Roma Beaumont and Gordon Reynell expertly drew the figures used in this book. Students taking our second year undergraduate course on "Environmental Management", and MSc students enrolled in "European Environmental Management", allowed us to test the main arguments of the book, and provided useful comments in the process. Parts of the manuscript were read by Keith Hoggart and Mark Mulligan and the book is the richer for their comments. Particular thanks need to be extended to John Thornes for his enthusiastic support of this project. The Department as a whole has been a highly conducive environment within which to work and all colleagues have provided support and encouragement along the way.

The constructive comments of David Alexander, Judith Rees and Tim O'Riordan on the draft manuscript are greatly appreciated. The support of Roger Jones as the editor of UCL Press has also been crucial to the successful completion of this book.

This book has been long in gestation and has involved substantial time commitment over several years. In this regard, the support of family and friends is greatly appreciated. The encouragement and patience of Olivia Wilson and Sara Thorne merit special acknowledgement. Olivia, in particular, has not only read the entire manuscript, but has also shown much forbearance with the inevitable day-to-day disruptions and discussions that are inevitably associated with the writing of a book of this kind.

LIST OF ABBREVIATIONS

CAP	Common Agricultural Policy
ELDCs	economically less developed countries
EM	environmental management
EMDCs	economically more developed countries
EU	European Union
GIS	geographical information systems
NGOs	non-governmental organizations
OPEC	Organization of Petroleum Exporting Countries
RTZ	Rio Tinto Zinc
TNCs	transnational corporations
UN	United Nations
WWW	World Wide Web

PART I

RE-EVALUATING ENVIRONMENTAL MANAGEMENT

This book promotes an inclusive understanding of environmental management (EM) as a multi-layered process. Part I sets out the analytical framework of the book. Chapter 1 relates how EM is both a process and a field of study, and explains who environmental managers are and what EM is. Chapter 2 takes up related analytical concerns by explaining the significance of uncertainty and predictability – concepts that are critical to the understanding of how environmental managers operate in multi-layered EM.

CHAPTER 1
Introduction

Never before in human history have environmental problems been such a central source of popular and scholarly concern. Issues such as tropical deforestation, ozone depletion, global warming and toxic waste disposal are given prominent coverage in the media, as the realization that environmental problems have attained global significance dawns on both politicians and public alike. Along with this growing interest has come an increased awareness of the central importance of understanding how human beings (mis)use the environment. Why this sudden interest? Ever since humans first started interacting with the environment, they have disrupted the natural environment. Although these processes were essentially local during most of human history, they have assumed global proportions in the twentieth century, and this globalization of environmental problems is likely to intensify in the twenty-first century.

Re-evaluating environmental management

As environmental problems have become increasingly apparent, so too has the need for a coordinated human response to these problems. This response is associated with the elevation of environmental management (EM) to a central concern in human–environment interaction at a hitherto unknown scale. Although the management of the environment has always been necessarily a human preoccupation, EM today is more important than ever. Indeed, and as Benton & Redclift (1994: 13) emphasize, "the management of the environment assumes urgency as we become more aware of what is going wrong in our relationship with the natural environment".

Paradoxically, research on EM has often been somewhat simplistic in its understanding both of what the process of EM is, and even which actors can be classified as environmental managers. This is not merely an issue of semantics; it goes to the heart of how EM is understood. Much of the literature has adopted a "state-centric" approach in that the subject matter is equated with the environmental policies and practices of the state. In some cases, this approach is explicit – MacNeill's (1971) book, for instance, was written at the behest of the Canadian government and it reflected the priorities and concerns of that government. In other cases, state centrism is implied through the choice of the topics examined, and the targeted audience. Atchia & Tropp (1995), for example, address a range of environmental problems and related management issues, but the underlying assumption throughout is that it is mainly the task of the state to resolve these problems. Many other works also reflect a traditional emphasis on the pivotal role of the state in EM (e.g. Petak 1980, 1981, Dorney 1987, Buckley 1991, Cooke & Doornkamp 1993). Even the Nath et al. (1993) major three-volume study on EM – which aims to develop a well rounded understanding of the subject – nonetheless tends to privilege topics, techniques and approaches most appropriately situated within a discussion of the state role in EM. As Sklair (1994: 206) argues, "the 'natural' approach to the whole world is state-centrist in a dual sense insofar as it emphasizes the role of the state and cognitively privileges the system of nation-states". Many non-state environmental managers are disempowered in the process.

This book takes issue with this traditional view of EM. It argues that EM needs to be understood in a more inclusive manner. Indeed, the very notion of an "environmental manager" requires reassessment to include non-state as well as state actors. As highlighted below, non-state actors such as farmers, hunter–gatherers, environmental non-governmental organizations (NGOs) or transnational corporations (TNCs), may be seen as environmental managers in the same way as state officials who "manage" the environment. This re-evaluation also requires a broader understanding of what is meant by "environmental policies", which are not necessarily synonymous with state policies.

Pioneering work by a growing group of scholars has begun to elaborate the bases of such a more inclusive understanding of EM. This work has specified the distinctive traits and management practices of diverse non-state environmental managers. For example, Welford (1996) has sought to delineate the EM role of many businesses, thereby acknowledging the growing influence of this type of actor in the EM process. In contrast, Wapner (1995) explores the increasingly significant EM role of transnational environmental NGOs, such as Greenpeace and Friends of the Earth. Rich (1994), meanwhile, offers a critical account of the indirect, but nonetheless often central, role of the World Bank (the leading international financial institution) in EM in many parts of the world. In addition, a growing literature highlights the EM role and practices of a diverse range of "grassroots" actors such as farmers, fishers, shifting cultivators, nomadic pastoralists and hunter–gatherers (e.g. Hong 1987, Bassett 1988, Denslow & Padoch 1988, Khasiani 1992, *The Ecologist* 1993, Fairlie 1995, Wilson 1996). These studies usefully highlight that EM is not the exlusive preserve of the state. Yet, they usually do not link empirical findings to broader questions surrounding the definition of EM. As a result, the significance of these findings is not yet fully appreciated in the field.

One of the aims of this book, therefore, is to attempt to draw together these diverse strands as the basis for arguing the need for a re-evaluation of traditional understandings of EM. It will emphasize the utility of an approach that understands the subject as a multi-layered process in which state and non-state environmental managers interact.

A second aim of the book is to suggest a possible analytical framework that facilitates understanding of how state and non-state environmental managers may interact with the environment, and with each other. The book suggests that, in a world of growing environmental problems, the essence of EM is the human attempt to promote predictability in a context of increasing uncertainty. It is argued that all policies and practices of environmental managers are related to the concepts of uncertainty and predictability as set out in Chapter 2.

This approach stresses the importance of an analytical understanding of human–environment interaction. As such, it stands in contrast to traditional approaches that often emphasize the description of environmental problems and their "solutions". The vast majority of books – even recent works by Castillon (1992), Nath et al. (1993), Pickering & Owen (1994) and O'Riordan (1995a) – are predominantly concerned with addressing such issues as global warming, water pollution or coastal degradation. In contrast, this book argues that EM may be more effectively understood through an analytical framework that encompasses themes relating notably to attitudes, policy, the market and politics. These themes need to be considered in the light of the multi-layered nature of EM.

To a certain extent, these issues have been addressed by a variety of authors (e.g. Mitchell 1989, Rees 1990, Porter & Brown 1991, Khasiani 1992, Born & Sonzogni 1995, Mather & Chapman 1995, Mitchell 1995). Yet, these studies have not sought to link their analyses to a re-evaluation of EM, whereas those who have attempted to "conceptualize" EM have failed to take into account the multi-layered nature of that process (e.g. Petak 1980, 1981, Bowonder 1987). The remainder of Part I of this book sets out an analytical framework that seeks to account for the differing research questions and objectives encompassed by the more inclusive understanding of EM adopted in this book.

What is environmental management?

What is the nature and meaning of EM? It is important to disentangle two different ways in which "environmental management" is used. On the one hand, EM can be thought of as a multi-layered process in which different types of environmental managers interact with the environment and with each other to pursue a livelihood. Critical here is how environmental managers seek predictability in their EM practices in a context of social and environmental uncertainty. On the other hand, EM can be understood as a field of study characterized by a set of concepts and approaches that interrelate in a distinctive way. The latter emphasizes the need for interdisciplinary understanding of human–environment interaction.

These two ways of understanding EM are clearly interrelated. A re-evaluation of one necessitates a reconsideration of the other. It is nonetheless the case that the logical place to begin is with analyzing EM as a process, and consequently this will be the major focus of this book. However, consideration is given both in this chapter and in Chapter 9 to reassessing EM as a field of study in light of the findings of this book.

Environmental management as a multi-layered process

Many scholars have considered the meaning of EM as a process. Yet, MacNeill (1971: 4) long ago noted that "discussions about the management of the environment are often plagued by misunderstandings and vagueness. This is due in no small part to the fact that the words environment and management mean quite different things to different people". Indeed, Harvey (1993: 2) notes that "words like 'nature' and 'environment' convey a commonality and universality of concern that is, precisely because of their ambiguity, open to a great diversity of interpretation". Notwithstanding these difficulties, to understand EM as multi-layered process is to clarify the meaning of "environmental management".

The literature is not lacking in proposed definitions in this regard. To begin with, scholars have sought to render precise the separate meanings of "environment" and "management". For example, Miller (1994: 731) defines environment as "all external conditions and factors, living and nonliving (chemicals and energy), that affect an organism or other specified system during its lifetime", whereas MacNeill (1971) includes even the built human environment in this definition. Similarly, "management" in relation to the environment has been defined in different ways. One view sees management as a means of allocating and conserving environmental resources (forests, minerals, etc.), whereas others emphasize management as a very structured process that "begins with goal setting and extends through the functions of information systems, research, planning, development, regulation and financing" (MacNeill 1971: 5). Bringing these two strands together, scholars have defined EM in various ways. Garlauskas's (1975: 190–91) definition is typical:

> environmental management is a creation of man [sic] . . . it centres on the activities of man and the relationships to the physical environment and the affected biological systems. The essence of environmental management is that, through a systematic analysis, understanding and control, it allows man to continue to evolve his technology without profoundly altering natural ecosystems.

As defined by Garlauskas (1975) and others (e.g. Petak 1980, 1981, Bowonder 1987), EM emerges as a process that tends to emphasize the application of science to specific environmental problems, usually under the auspices of the state (see also Schafer & Davis 1989, Atchia & Tropp 1995). Conspicuously absent from such traditional accounts is any sense of the complex political, economic and social interactions of different types of actors pursuing EM. As such, it can be argued that this understanding constitutes an inadequate basis for conceiving of EM as a multi-layered process.

What then are the elements of a more inclusive definition of EM as a process? Three points need to be noted here. First, such a definition must render explicit the multi-layered nature of EM as a process. This entails acknowledging not only the role of the state, but also specifying the activities of diverse non-state environmental managers. Secondly, the definition must avoid the trap of equating EM with professional training and expertise. Just as non-state environmental managers need to be incorporated in an understanding of EM, so too room must be made for an appreciation of "non-professional" and "non-scientific" approaches to managing the environment. The latter approaches typically do not figure in many state-centric accounts. Rather, the assumption is often that environmental managers can only be trained professionals or "team leaders" (e.g. Petak 1980: 290). This implies that no other individuals interacting with the environment can be classified as environmental managers. Thirdly, an inclusive definition must also render explicit the central predicament of all environmental managers, namely the quest for predictability in a context of increasing social and environmental uncertainty. The following definition, therefore, clarifies what EM is and who environmental managers are in light of these three concerns:

> Environmental management can be defined as a multi-layered process associated with the interaction of state and non-state environmental managers with the environment and with each other. Environmental managers are those whose livelihood is primarily dependent on the application of skill in the active and self-conscious, direct or indirect, manipulation of the environment with the aim of enhancing predictability in a context of social and environmental uncertainty.

As understood here, "environment" relates to, but need not be confined to, physical and biological processes. As Chapter 8 illustrates with reference to "virtual EM", there is a potential need to adopt a more inclusive approach towards "environment" that captures the growing importance of "imaginary environments" in EM. The primary meaning attached to environment in this book is, nonetheless, that of the physical environment.

The definition set out above demands a wider definition of which actors are environmental managers. It is, therefore, imperative to explore the possible different types of environmental managers that may interact in multi-layered EM. Table 1.1 sets out in a simplified form the main types of environmental managers mentioned in this book. The objective of the table is not to describe every conceivable type, but rather to illustrate the analytical implications of a more inclusive definition.

The role of the state as a type of environmental manager needs to be

Table 1.1 Key types of environmental managers.

Environmental managers	Examples	Level of environmental interaction
State	Department of the Environment Department of Forestry Ministry of Agriculture	Actively and self-consciously manage the environment at the local, national and global levels
Environmental NGOs	Greenpeace Friends of the Earth Haribon (Philippines)	Active and self-conscious role in influencing decisions about EM at the local, national and global levels
TNCs	Rio Tinto Zinc Matsui Siemens	Actively and self-consciously manage the environment at the local, national and global levels
International financial institutions	World Bank International Monetary Fund Asian Development Bank	Active and self-conscious role in influencing decisions about EM at the local, national and global levels
Farmers, fishers, nomadic pastoralists, shifting cultivators	Farmers in the UK Spanish fishers Buroro nomads (Africa)	Actively and self-consciously manage the environment at the local and regional levels
Hunter–gatherers	Penan (Malaysia) Yanomami (Brazil)	Actively and self-consciously manage the environment at the local level

Source: authors.

acknowledged, but, as Table 1.1. highlights, many other non-state environmental managers can be identified. For example, the prominence of environmental NGOs, TNCs and international financial institutions at different levels of environmental interaction is incorporated in this inclusive understanding. That understanding also encompasses a range of predominantly local-level ("grassroots") environmental managers: farmers, fishers, nomadic pastoralists, shifting cultivators and hunter–gatherers: EM is a process not exclusive to large national and international environmental actors.

A further benefit of the inclusive understanding of EM relates to the question of indigenous versus Western positivist[1] knowledge in EM. The spread of such Western science around the world has often been based on a rejection of indigenous knowledge upon which local EM has long been based (Adas 1989, Redclift & Woodgate 1994, Pretty 1995). The inclusive approach adopted here will help to reverse this complex trend by emphasizing the nature of knowledge construction.

The different types of environmental managers involved in a multi-layered EM are often different in terms of their environmental impact, motivations or

1. Based on a quantitative and rational understanding of the environment.

interests. However, they all can be considered as "environmental managers" insofar as all their livelihoods are primarily dependent on the application of skill in the active and self-conscious manipulation of the environment (see the definition quoted above). Yet, there are considerable differences among these types of actors. Whereas farmers, hunter–gatherers or many TNCs and state agencies directly manipulate the environment (e.g. agriculture, forestry, gathering of forest resources, mining), environmental NGOs and international financial institutions derive their livelihoods (in part or in total) from the indirect manipulation of the environment (e.g. information campaigns, loan provision). The similarities and differences between these different types of environmental managers are a recurring theme throughout this book.

It needs to be reiterated that the types of environmental managers set out in Table 1.1, and discussed throughout the book, do not cover all possibilities. There are certainly other types of environmental managers that could be mentioned. For example, professional environmental consultants can be classified as environmental managers as they indirectly manipulate the environment as per the definition noted above. Similarly, media professionals specializing on EM issues also can be seen as "environmental managers". In effect, the latter may play a similarly indirect role in multi-layered EM to environmental NGOs or international financial institutions. In a similar vein, many small to medium-size businesses interact with the environment, in keeping with the above definition (cf. Welford 1996). However, for the purposes of this book, the discussion will focus on the types set out in Table 1.1.

The adoption of a more inclusive understanding of EM inevitably raises the vexed question of who is and who is not an environmental manager.

To this end, it is useful to distinguish between "environmental managers" and "environmental users". All human beings are environmental users, as all people interact with the environment in the pursuit of their needs and wants. However, not all environmental users are environmental managers. The appropriate definition of EM and environmental managers must be situated somewhere between these two extremes. Critical here is the definition of EM set out above. All those actors who do not fall within the terms of the definition will be referred to in this book as "environmental users".

Two examples suffice to illustrate the point. Individuals such as stockbrokers or musicians whose livelihood is not "primarily dependent on the application of skill in the active and self-conscious manipulation of the environment" are not environmental managers. For similar reasons, those individuals who may have an appreciable impact on the environment through consumption or leisure activities, but who do not derive a livelihood from these activities in the process (e.g. car users, wilderness visitors), must also be excluded from the category "environmental manager". The point here is not to suggest that these or other environmental users do not have a potentially important impact on the environment. To the contrary, they often are linked to many of the pressing environmental problems confronting environmental managers today (see Ch. 3). Rather, the discussion seeks to clarify the distinguishing traits of environmental managers, as well as the opportunities and constraints that they face in multi-layered EM. Thus, environmental users figure in this book only insofar as they have a bearing on the policies and practices of environmental managers.

A different consideration is associated with the complex interests and motivations that relate to each type of environmental manager referred to in Table 1.1. It is straightforward to appreciate the role of a farmer or hunter–gatherer as an environmental manager. Less clear is the precise delineation of EM responsibilities and practices within more "complex" types of environmental managers, such as states or TNCs. Once again, the yardstick to be used in evaluating whether an individual within a TNC or the state can be classified as an "environmental manager" is whether the activities of these individuals conform to the above definition (i.e. livelihood primarily dependent on the self-conscious manipulation of the environment).

Reference to the state as an environmental manager, for example, denotes those individuals whose primary livelihood relates to EM through line agencies (e.g. forestry departments, agriculture ministries) or parastatal[1] organizations (e.g. state energy corporations, state forestry corporations). As individuals who hold ultimate responsibility for the direction of state activities (including EM) many political leaders also are environmental managers. Thus, although it might be convenient to speak of "state EM" in this book, such management in practice encompasses a multitude of different EM strategies undertaken by different state agencies and individuals.

Other types of environmental managers, such as international financial institutions, TNCs or environmental NGOs, may also display much internal complexity as organizations. As with the state, there is usually a division of responsibilities with regard to EM. In TNCs, for example, responsibility often resides with especially designated environmental managers who liaise between the different decision-making levels within a corporation. That said, ultimate management responsibility typically rests with the chief executive officer and board members, who must weigh up EM considerations in line with the company's other priorities (Welford 1996).

In speaking of "types" of environmental managers, it is also important to recognize that in some cases not every member of that category can be classed as an environmental manager. The case of TNCs (but also medium-size to small businesses) illustrates this point. TNCs such as Weyerhäuser, Rio Tinto Zinc (RTZ) and Siemens, who are involved in the active and self-conscious direct manipulation of the environment through such activities as logging, mining or waste management, clearly fall within the definition of environmental managers. Similarly, TNCs such as Jaakko Poyry or the Sandwell Company (environmental consultants) also actively and self-consciously manipulate the

1. Organizations closely related to the state.

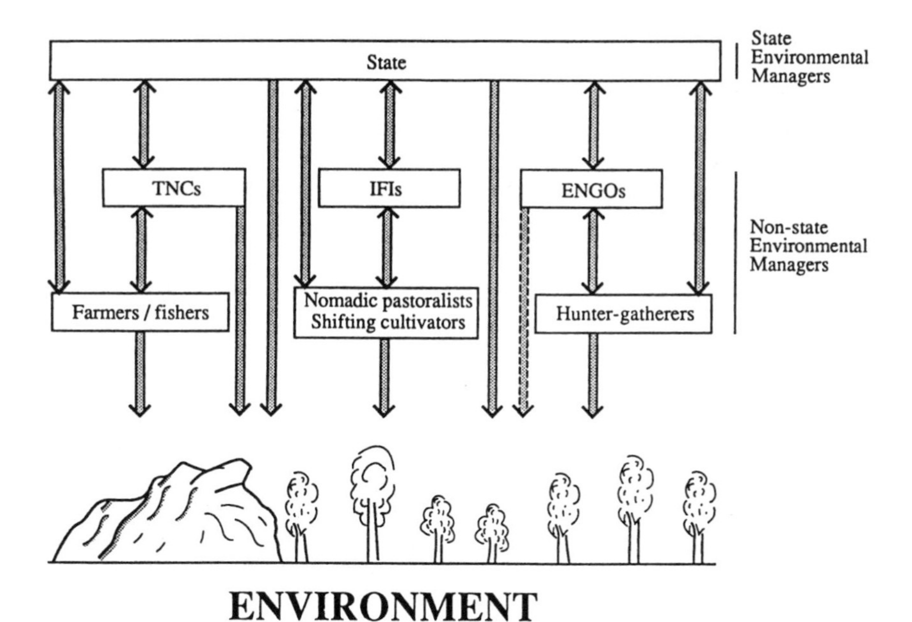

Figure 1.1 Environmental management as a multi-layered process.

environment indirectly through advising other environmental managers and are, therefore, also environmental managers. In contrast, TNCs not involved in the active and self-conscious, direct or indirect, manipulation of the environment are not environmental managers but environmental users.

Central to an inclusive understanding is the idea that EM is a multi-layered process. Figure 1.1 illustrates that process, depicting the interrelationships of the different types of environmental managers highlighted in Table 1.1. It is important to note that the figure does not aim to show all the possible interactions between environmental managers, but seeks to illustrate selected examples to illustrate broader principles. Understanding EM as a multi-layered process is inextricably interlinked with the adoption of an inclusive understanding of EM.

Three general points about Figure 1.1 need to be made. First, it reiterates that EM encompasses the management practices of both state and non-state environmental managers. Although the role of the state is often crucial to the overall process of EM, the significant role of diverse non-state environmental managers should not be neglected.

Secondly, Figure 1.1 highlights that EM usually involves the interaction of environmental managers directly with the environment, but that such direct interaction is not a prerequisite for status as an environmental manager. Thus, international financial institutions exclusively, and environmental NGOs predominantly, are involved in the indirect manipulation of the environment mainly through their relationships with other environmental managers. Indeed, an increasingly important component of state EM is similarly indirect in nature (e.g. regulations).

Thirdly, Figure 1.1 also highlights the importance of understanding EM in terms of the interaction of different types of environmental managers with each other. This point emphasizes that environmental managers do not operate in isolation from each other, and that their practices are often strongly influenced by the policies and practices of other environmental managers.

The implications of these points, and of multi-layered EM generally, are returned to throughout this book. For example, the issue of the growing human impact on the environment (Ch. 3), and the likely social and technological impact in future EM (Ch. 8), are grounded in an appreciation of the different effects of such impacts on environmental managers operating in multi-layered EM. However, it is through the exploration of the role of worldviews, attitudes and discourses (Ch. 4), politics (Ch. 5), markets (Ch. 6), and policies (Ch. 7) that the utility of re-evaluating EM in terms of a multi-layered process is best illustrated.

It is useful, nonetheless, to illustrate briefly here the importance of understanding EM as a multi-layered process. It is possible to explore the EM practices of local-level environmental managers such as farmers or hunter–gatherers in order to develop a sense of how EM is practised by these actors. Yet, their EM practices rarely occur today in isolation from the policies and practices of other environmental managers operating within multi-layered EM. Notable in this regard is that these grassroots actors often have to take into account the role and interests of state environmental managers. In some cases, this process may involve adapting practices to state specifications (e.g. certain environmental regulations or laws). For hunter–gatherers or shifting cultivators, conflict may be endemic to the relationship between these types of environmental managers and the state, because of the incompatibility of their EM practices with those of state environmental managers. Further complexity in multi-layered EM is derived from the growing importance of TNCs, international financial institutions and environmental NGOs who, through their practices often circumvent, if not directly subvert, state EM. In this process, EM needs to be seen as the interaction of different types of environmental managers in contexts characterized by conflict and cooperation (Ch. 5).

To re-evaluate EM is, thus, to reconsider what EM is, and who environmental managers are. Two further points need to be emphasized in considering the implications of this endeavour. First, the definition used here does not assume that the process of EM necessarily results in sustainable use of the environment. Hence, it is possible – and indeed, as this book discusses, it is often the case – that EM contributes to environmental degradation. It is, therefore, necessary to differentiate between sustainable and unsustainable EM (see Ch. 2).

Secondly, the new definition has important practical political implications. The state-centric perspective has sanctioned an approach in which state environmental managers are seen as the "natural" leaders, whereas non-state actors have been relegated to a subsidiary role (Sewell 1973, Freeman & Frey 1986). This approach has been rightly criticized in recent years (Sachs 1993). As Benton (1994: 37) argues, "the 'environmental management' perspective has tended to under-theorize the social, legal and political processes of environmental regulation themselves". The problem is not so much with the concept of EM and environmental managers per se, but with the hitherto exclusive nature of how this concept has been understood. This book, which contributes to a reassessment of the relative importance in EM of state and non-state environmental managers, emphasizes the central political importance of the latter. That recognition is not exclusively a scholarly matter, but also necessitates the reconsideration of the role of non-state environmental managers in the design and implementation of EM policies and practices.

This section has argued that a re-evaluation of EM is an essential part of a wider effort to develop a well rounded understanding of how humans manage the environment. This re-evaluation also has implications for the ways in which EM is thought of as a field of study.

Environmental management as a field of study

Today, EM is what Tansley, writing in 1904 with regard to ecology, called a "fashionable study" (in Park 1980: 33). However, just as EM has been thought of as a process that is mainly the concern of state professionals, so too the literature on EM has often been technical and written by, and predominantly for, professionals (e.g. Dorney 1987, Buckley 1991). As a result, EM is perceived from outside the field as a relatively obscure and jargon-laden field of study (Sachs 1993).

The origins of EM as a field of study lie in the growing concern about environmental degradation in the 1960s and early 1970s. For the first time, there was a recognition that humankind had the ability to damage the Earth's environment irreparably. From the start, concern in EM was linked to an attempt by scholars to recommend to governments ways in which to modify EM practices. Hence, EM has been perceived within the social and natural sciences as a predominantly applied field of study. In 1971, for example, an important book by Jim MacNeill explored basic precepts of EM in the context of advice and recommendations to the Canadian government. In the same year, O'Riordan (1971) sought more specifically to define the parameters of "resource management" for both the scholarly and policy-making communities. However, it was the launch in 1973 in the UK of the *Journal of Environmental Management* that signalled the growing coherence and confidence of this emerging field (Jeffers 1973). Subsequently, a journal entitled *Environmental Management* was established in the USA in 1976, also dedicated to the discussion of "professional" EM concerns of the scientific and policy-making communities (DeSanto 1976,

Sandhu 1977, Alexander 1985). Recently, for example, this process has also been reflected in both the relaunch of the *Journal of Environmental Planning and Management* under its new and more explicitly EM-based title, or the ambitious launch of five interlinked new journals by the publisher John Wiley orientated around the theme of EM.

Many studies have sought to elaborate the field of EM, typically emphasizing its complex nature (e.g. Petak 1981). For example, Park (1980) sought to integrate EM with ecology from a geographical perspective, an endeavour taken further by Cooke & Doornkamp (1993), but with an emphasis on linking EM specifically to geomorphology. Other work has sought to situate EM in an historical context (e.g. Powell 1976), and perhaps most ambitiously with environmental science (O'Riordan 1995a). However, many books have reiterated the link between EM and the role of state professionals (e.g. Dorney 1987, Mitchell 1989, Buckley 1991). Thus, research has simultaneously sought to elaborate the disciplinary interactions and policy-making implications of EM. Such research notwithstanding, MacNeill's (1971: 6) early claim that "it is difficult to achieve consensus on what the subject of 'environmental management' should and should not include" still rings true today. As Nath et al. (1993: 19) observe, although EM "has a common sense ring to it . . . what it entails in terms of philosophy, methodology and content is far from obvious". Many have shied away from responding to the challenge of elaborating EM (Moffat 1990). For example, Petak's (1981: 213) self-described conceptualization of EM was nevertheless not intended to "engage in a highly detailed analysis, which would be the subject for a major treatise in the field of environmental management".

To a large extent, this problem reflects confusion over the disciplinary basis of EM. Over the years, EM has been variously associated with such disciplines as biology, ecology, economics and geography. Such associations have reflected more about the disciplinary origins of the authors than the intellectual merits of locating EM in one discipline or another. As a result, EM has been defined in relation to differing disciplines, but in such a way that no overall coherence has emerged.

Yet, certain key themes in EM would tend to suggest that this field is more closely affiliated to some disciplines than others. As MacNeill (1971) pointed out, space is a crucial dimension of EM in that both environmental problems and proposed solutions are spatially defined. Concurrently, O'Riordan (1971) argued that resource use and allocation are essential to EM. Both MacNeill and O'Riordan separately have thereby emphasized the central importance of geography to EM (see also Birch 1973). Subsequent work by geographers such as Park (1980), Johnston (1983), Rees (1990), Trudgill (1991), Cooke & Doornkamp (1993) and Mitchell (1995) have reiterated this specific disciplinary affiliation.

Scholars have emphasized other disciplinary links. For example, Mitchell (1989) and Rees (1990) have further highlighted the economic dimensions to EM. In this context, since the mid-1980s research in environmental economics, notably work by Barbier (1989), Pearce et al. (1989) and Schramm & Warford

(1989), has emphasized the possible affinities between economics and EM. Others have made the connection between EM and planning studies (Birch 1973, Allison 1974). Once again, much of this work is written from a state-centric perspective with a view to advising state professionals about appropriate state EM practices.

One of the most significant disciplinary interactions relates to the link between EM and environmental science. The latter, which combines elements from such natural scientific disciplines as biology, chemistry and physics, developed separately around the same time as EM (Botkin & Keller 1995, Ennos & Bailey 1995, O'Riordan 1995a). Yet, despite many shared concerns, EM and environmental science have followed different development paths. Indeed, the split between environmental science (a largely natural scientific field of study) and EM (a predominantly social scientific field of study) has been such that it is only recently that scholars have attempted to relate the two in a systematic manner. Yet, as O'Riordan (1995a) illustrates, although there are many important possible interactions between the two fields of study, no work to date has been able to bridge successfully the division that exists between them. In effect, whereas environmental science is predominantly concerned to explain the physical bases of environmental change, EM is mainly focused on understanding how humans interact with the environment. The latter takes largely as given the explanations of the physical environment that the former seeks to provide.

Figure 1.2 suggests a way of understanding disciplinary influences on EM based on EM as a multi-layered process. Without being exhaustive, it emphasizes the complex ways in which EM is related to the social sciences, and to a much lesser extent to the natural sciences. Indeed, and as Benton & Redclift (1994: 11) point out, "research on the environment occurs at different points of convergence between disciplines". This has the result that disciplinary connections are becoming increasingly complex, as solutions are sought from a wide array of disciplines in multi-layered EM.

This point is reflected in the disciplinary influences on EM. As Figure 1.2 illustrates, EM largely draws intellectual sustenance from disciplines and sub-disciplines within the social sciences. To the extent that the natural sciences are an influence on EM, relevant disciplines such as ecology, biology and chemistry typically become relevant for EM under the auspices of environmental science. The overall influence of the natural sciences on EM is, nonetheless, limited. A similar comment can be made with regard to the influence of physical geography (as the natural scientific component to geography) on EM.

A more complex set of disciplinary influences infuse the relationship between the social sciences and EM. This follows logically from conceiving of EM as a multi-layered process. To begin with, certain disciplines are more important than others. For example, politics, economics or anthropology play a particularly critical role, whereas philosophy, history and psychology tend to be a secondary influence on EM. However, as Figure 1.2 also illustrates, the

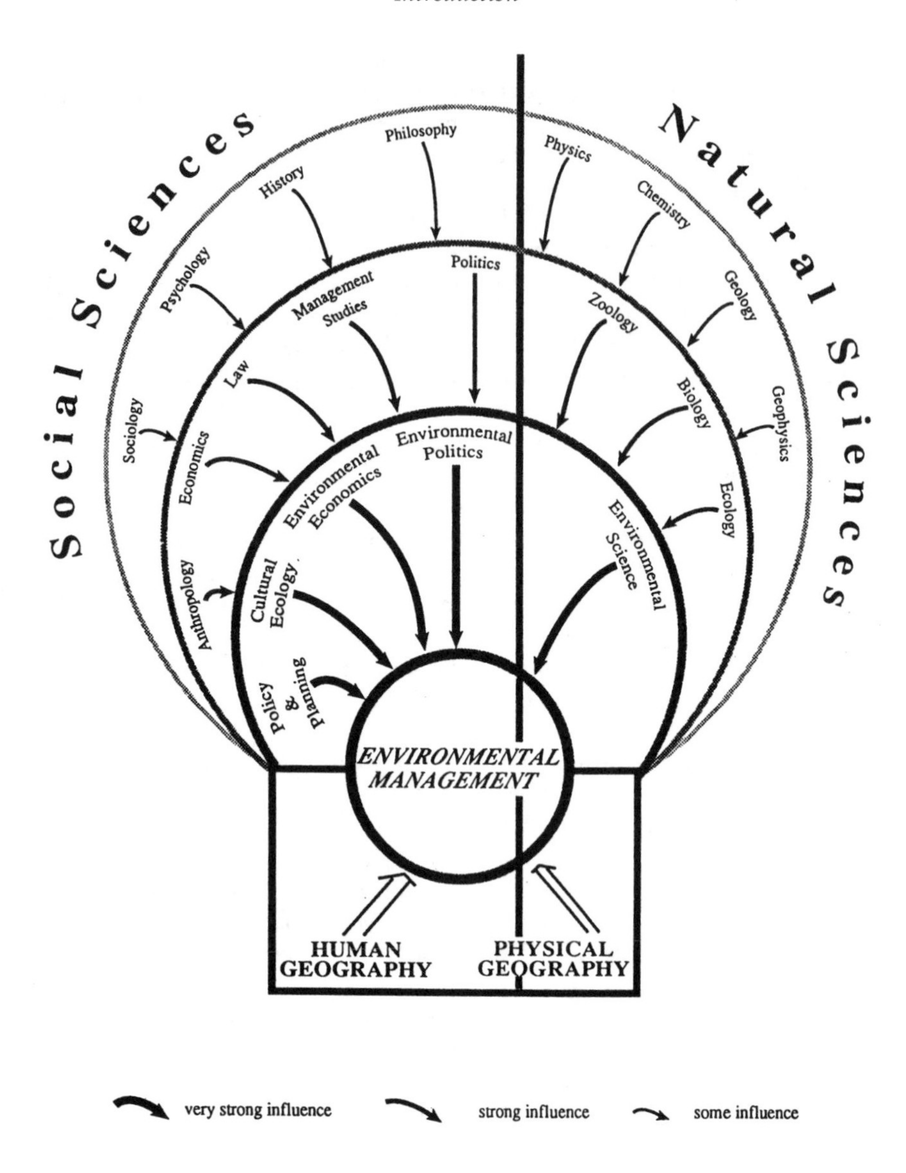

Figure 1.2 Disciplinary influences on environmental management.

main link between the social sciences and EM is through a handful of subdisciplines. Thus, environmental politics (politics), environmental economics (economics) and cultural ecology (anthropology) play a pivotal role in influencing research themes in a re-evaluated EM (see also Ch. 9). Figure 1.2 reiterates the point made above that the closest affiliation of EM is with geography, especially human geography. The focus of this discipline on issues of space, scale, and human–environment interaction generally is indicative of the "natural" relationship between geography and EM.

The relationship between EM and resource management warrants separate treatment. The latter has not been included in Figure 1.2 because it can be treated as a subfield of EM. All work in resource management is also EM, but the reverse is not necessarily the case. EM is concerned with the management dimensions of the totality of human–environment interaction, and not simply those dimensions pertaining to resource use (cf. Mitchell 1989, 1995, Rees 1990).

It is important to note that EM operates at the intersection of a range of disciplines and subdisciplines (see Fig. 1.2). In light of the re-evaluation of EM advocated in this book, disciplinary boundaries may represent more of an obstacle than an opportunity in the development of this field (see Ch. 9). Disciplinary boundaries often have served to channel, and in the process limit, intellectual inquiry. Perhaps more than any other subject, EM necessitates the interaction of different disciplinary approaches as part of a broader quest to develop an integrated understanding of human–environment interaction. In this light, Harvey (1993: 38) argues that "some kind of transdisciplinary language is required to better represent and resolve ecological problems", a point that is essential in EM as it draws upon diverse disciplinary influences to explain and understand the management of the environment.

Insofar as the reassessment of EM necessitates a different perspective on human–environment interaction, a question arises as to whether current developments in the field signal a "paradigm[1] shift" (Kuhn 1970). Devised as a means to explain changes in scientific inquiry, Kuhn's notion of the paradigm may be important here in that it highlights how research is socially constructed. Although this notion has been criticized for being ambiguous and rigid (Stoddart 1981), for a subject as wide-ranging as EM it may be useful for understanding the development of social and natural scientific inquiry on the environment (Colby 1991). This book does not seek to answer this question but, as Chapter 9 briefly illustrates, a reassessment of EM may contribute to developments that one day may lead to a paradigm shift.

Conclusion

This book contributes to a reassessment of EM. It argues that there is a need for a more inclusive understanding of the subject matter that incorporates the EM role of state and non-state actors. A definition was introduced based on the

---

1. A paradigm comprises researchers committed to the same rules and standards for scientific practice as a prerequisite for the genesis and continuation of a particular research tradition (Kuhn 1970: 11).

idea of EM as a multi-layered process. The reassessment advocated in this book has important implications for understanding EM both as a process and as a field of study. The book integrates the analysis of state and non-state actors in order to develop a more inclusive understanding of the process of EM. It further argues that the field needs to reflect this broader understanding, notably through a different mix of disciplinary interactions emphasizing social scientific enquiry.

The measures discussed in this chapter only contribute partially to a reassessment of EM. What still needs to be explored is the way in which all environmental managers pursue predictability through their practices, so as to reduce social and environmental uncertainty. Chapter 2 examines the concepts of uncertainty and predictability as part of the further elaboration of this book's analytical framework centred on the idea of EM as a multi-layered process.

CHAPTER 2
Uncertainty and predictability

To understand the policies and practices of environmental managers, it is crucial to appreciate how they seek predictability in the face of social and environmental uncertainty. In this chapter we will explore in greater detail the dimensions and implications of these two concepts.

Uncertainty

Social and environmental uncertainty is the greatest problem facing environmental managers in multi-layered EM. Such uncertainty has certainly always been part of human–environment interaction, but in recent years it has become more important (Beck 1992, O'Riordan 1995a). As Petak (1981: 214) highlights, EM "must deal with a large range of uncertainties . . . In this context, management must work to reduce uncertainty, while searching for the flexibility necessary to respond to changing social and political values and demands". The following discussion is mainly concerned to elucidate the specific types of uncertainty that environmental managers may confront. First, though, it is important to clarify general issues surrounding the concept of uncertainty.

The concept of uncertainty is useful as a means to explain the general social and environmental conditions that confront all environmental managers. It differs from the concept of "risk", as used notably in many traditional accounts in EM. The concept of risk is usually defined with reference to "terms such as probability, likelihood or chance which imply the likelihood of something occurring" (Gerrard 1995: 301). The emphasis is on risk assessment that involves "using data, assumptions, and models to estimate the probability of harm to human health or to the environment that may result from exposures to specific hazards" (Miller 1994: 533; see also Brown et al. 1995, Adams 1995). It is not difficult to see why this concept has long figured in traditional accounts of EM (e.g. Foran & Stafford Smith 1991, DuVair & Loomis 1993). Its emphasis on positivist science to quantify probabilities of harm fits well with the broader problem-solving ethos of those accounts.

Yet, just as that ethos was found to be wanting in terms of providing a well rounded understanding of EM as a multi-layered process, so too the utility of this understanding of risk may be questioned as a means to explain the social and environmental conditions that may confront all environmental managers. Within the social science literature, this traditional understanding of risk has been challenged in recent years, notably in the work of Beck (1992). Beck's argument will be explored in more detail below, but what needs to be noted here is this author's emphasis that risk is a cultural, not simply a technical, phenomenon. This broader definition of risk rescues the concept from the narrow interpretation usually adopted. Yet, for the purposes of this book even this expanded notion of risk is not suitable as a description of the totality of social and environmental conditions that influence EM policies and practices, because it does not address all facets of the problem. Rather, the concept of "uncertainty" will be used to denote those conditions pertinent to decision-making in EM.

At a general level, uncertainty needs to be related to questions of scale and time. In preindustrial times, many environmental managers were preoccupied predominantly with subsistence-orientated EM, whereas in the industrial era those issues have encompassed a wider range of concerns linked to intensified human use of the environment and the globalized capitalist system (see Ch. 3). Uncertainty is also expressed at different scales. Issues pertaining to deforestation, soil erosion and unequal power relations are often at the heart of uncertainty at the local level. In contrast, global environmental problems (e.g. global warming or ozone depletion) and international conflict among certain types of environmental managers reflect uncertainty at a global level. However, such uncertainty cannot be compartmentalized neatly into local and global scales. Indeed, as the issue of the link between tropical deforestation and global warming illustrates, uncertainties at one scale may serve to reinforce uncertainties at another scale (Clayton 1995).

In a similar manner, uncertainty can be differentiated according to its dimensions. These include environmental, sociocultural, political and market-related uncertainties. A brief exploration of these dimensions will help to clarify the meaning and significance of the concept of uncertainty.

Environmental uncertainty is perhaps the most obvious problem confronting environmental managers. Such uncertainty, in turn, derives from problems associated with both environmental degradation and the possible "chaotic" characteristics of natural environmental processes. Environmental degradation relates to the adverse consequences of human-induced environmental change. Table 2.1 sets out the multiple dimensions to such degradation and shows that environmental degradation includes "taking" from, "adding" to, "replacing" and "modifying" the environment. The table illustrates that environmental degradation is not only a resource scarcity issue, but also encompasses the full range and impact of human–environment interactions.

Environmental degradation resulting from taking from the environment includes the depletion of various natural resources (Rees 1990). The felling of the world's old growth forests for timber and other uses, soil erosion and wildlife depletion are good examples (Williams 1989, Rush 1991). Environmental

Table 2.1 The multiple dimensions of environmental degradation.

	Process	Example	Impact
Taking	Taking resources	Soil erosion	Depletion
Adding	Adding pollution	Sewage	Poisoning
Replacing	Replacing natural with anthropogenic environments	Plantations	Simplification
Modifying	Genetic manipulation	High-yielding crop varieties	Uniformity

Source: adapted from Dunlap 1993: 711–12.

degradation can also be caused by adding to the environment the by-products of human activity such as toxic wastes, sewage and CO_2 emissions that "poison" land, water or air. Increased environmental uncertainty here is linked to global warming and ozone depletion (Clayton 1995). Degradation also results from replacing natural environments with simplified anthropogenic environments. For example, old-growth forests are replaced by timber plantations, resulting in biodiversity loss (Sargent & Bass 1992, Marchak 1995). Finally, environmental degradation may also be linked to modifying the environment through the introduction of genetically modified species into the natural environment, thereby again reducing biodiversity (Kloppenburg 1988).

Table 2.1, thus, highlights the multiple dimensions of environmental degradation. Such degradation interacts in complex ways and has increased over time, thereby enhancing environmental uncertainty (see Ch. 3). Such uncertainty is also compounded by poor understanding of environmental degradation (Reckow 1994). Many environmental issues remain poorly understood because of insufficient or unreliable data. Further, perceptions about the meaning of "degradation" itself vary among environmental managers, notably in keeping with different cultural contexts and discourses (Fairhead & Leach 1996; see Ch. 4).

Environmental uncertainty may also be inherent in natural environmental processes themselves. As O'Riordan (1995a: 8) emphasizes, "there is a school of thought which claims that certain natural processes are indefinable and indeterminate because they operate in mysterious ways that can never be fully understood". A case in point is the discussion suggesting that ecological processes are inherently unpredictable (Gleick 1987, Worster 1993). What has become known as "chaos theory" suggests that, contrary to conventional wisdom, ecological processes do not tend towards equilibrium and order (Prigogine & Stangers 1985). Although the traditional view has been that any disturbance to an ecosystem will lead to adjustments within that system that aim to restore the status quo ante (Odum 1971), chaos theory asserts that the "order of things" is a randomness that scientists are unable to understand fully (Worster 1993, Zimmerer 1994). Debate continues to swirl around these competing perspectives on ecological processes, but the possible implications for

EM may be briefly noted. The emphasis of the traditional view on ecosystem stability has lent itself to positivist approaches in EM, by suggesting that environmental degradation can be measured, monitored and rectified through the application of scientific knowledge. Yet if, as chaos theorists suggest, such stability is an illusion, then EM practices predicated on the idea of stability are misguided, if not even detrimental to environmental wellbeing in a chaotic world.

The debate surrounding the dynamics of ecological processes highlights usefully some of the potential dangers associated with adopting an approach that does not acknowledge the complexity of both ecological processes and the human ability to understand these processes. Further, it is a vivid reminder of the possible limitations of "professional science" as applied to EM. The discussion illustrates above all that ecological processes are not a "given", and to the extent that they may be inherently unpredictable, environmental uncertainty in EM cannot be avoided. Indeed, when added to the uncertainty associated with human-induced environmental degradation, doubts surrounding scientific knowledge and control of ecological processes underscore why environmental uncertainty usually plays a major role in the policies and practices of environmental managers.

Yet, for environmental managers, uncertainty is associated as much with human behaviour as it is with environmental processes. Clearly, the two are linked. Environmental degradation is one outcome of human–environment interaction, but, just as uncertainty may be at the heart of ecological processes, so too uncertainty is embedded in human societies and is associated notably with sociocultural, political and market activities.

Sociocultural uncertainty emerges from complex individual and group worldviews, attitudes, discourses and behaviour in society. As noted above, uncertainty is often linked to environmental issues. Yet, such uncertainty is not a given, but is rather interpreted subjectively by individuals and groups, depending on how they perceive their interaction with the environment. What may be an environmental problem to some, may not be a problem to others (Blaikie & Brookfield 1987). Thus, environmental attitudes themselves can be a basis for uncertainty in EM. Those attitudes, in turn, are linked to contrasting environmental worldviews that are often the basis of strikingly different interpretations of what is "appropriate" EM. Further, the articulation of attitudes and worldviews through selected discourses about the causation and meaning of environmental problems often exacerbates the sociocultural uncertainty confronting environmental managers (see Ch. 4).

Sociocultural uncertainty is also associated with broader trends in society that may be changing human–environment interactions irrevocably. In this context, the notion of "risk society" developed by Beck (1992) provides a useful means to understand such broad changes (see also Lash et al. 1996). As used by Beck, "risk" is understood in relation to sociocultural upheavals of post-industrial societies. Specifically, Beck argues that a by-product of industrialization has been the generation of unwanted risks that have accumulated to such an extent that they threaten the survival of humanity. In the case of certain risks, associated notably with nuclear waste and pesticide use, the threat is a general and "invisible" one, having both short and long-term adverse effects on the environment. These risks "induce systematic and often irreversible harm, generally remain invincible, are based on causal interpretations, and thus initially only exist in terms of the . . . knowledge about them" (Beck 1992: 22–3). Beck certainly acknowledges that these risks may be unevenly distributed in society, with the poor typically more exposed to them than the rich, reflecting unequal power relations in society (cf. Bullard 1993; see below). Yet, the main thrust of his argument is that generalized risks exist and are increasingly shaping the character of post-industrial societies today. What is important to emphasize here is that this qualitative series of sociocultural changes – associated notably with the transition from "industrial" to "risk societies" – compounds pre-existing sociocultural uncertainties rooted in worldviews, attitudes and discourses. The emerging picture is one of increasing sociocultural uncertainty in which environmental managers must seek to pursue their policies and practices.

Political uncertainty is another factor confronting environmental managers, in that constraints and opportunities in EM are often associated with political processes. Those processes need to be understood broadly as a reflection of the unequal power relations that structure human–environment interactions. This has important implications for the distribution of uncertainty in society. Specifically, a central characteristic of multi-layered EM is that social and environmental uncertainty is unequally distributed among environmental managers. Indeed, power may in part be about the ability of one environmental manager to shift such uncertainty to another environmental manager. Poor environmental managers are often more susceptible to environmental degradation and "hazards" because of their marginal political and economic status in society (Watts 1983b, Perry & Nelson 1991, Watts & Bohle 1993, Blaikie et al. 1994). That status is often a corollary of the ability of wealthy and powerful environmental managers to influence EM decision-making either through the state or through interactions with other non-state environmental managers, so as to minimize the social and environmental uncertainty they face. More powerful environmental managers are better placed than their weaker counterparts to use the state's legal and coercive abilities to facilitate their control of many environmental resources and labour. The result for weaker environmental managers is usually a context in which political uncertainty looms large in the guise of social repression and reduced access to essential environmental resources.

It is the poor and the weak who suffer disproportionately from political uncertainty. Yet, however unequal, power relations are ever in flux, and no environmental manager is able to control political processes permanently. Hence, all environmental managers inevitably face political uncertainty to a greater or lesser extent. Large businesses, for instance, may often be able to influence political processes so as to facilitate production and profit maximization, but even these firms must always be wary of social changes that might result in increased regulation that adversely affects profits. Such uncertainty is often reflected in the policies and practices of different environmental managers.

Market uncertainty adds a different, but no less important, set of potential constraints for environmental managers. The market serves to link together consumers and producers at various scales, and as such is an important framework within which environmental managers pursue their livelihoods. Yet, if participation in the market may be an opportunity, it also often represents an important source of uncertainty. Except in rare cases, the market is not controlled by any one actor or group of actors, but reflects the complex interactions of diverse interests (Rees 1990, Eckersley 1996a). Uncertainty for environmental managers derives from fluctuations in the price, quantity, and quality of market goods that often leave individual environmental managers unsure of their livelihoods. It is certainly the case that market fluctuations may impact less severely on economically powerful environmental managers (e.g. TNCs) than on their less powerful counterparts.

The preceding discussion has briefly illustrated the environmental, sociocultural, political and market uncertainties that may confront environmental managers. Such uncertainties may act often in combination, intensifying the overall uncertainty faced by some environmental managers. It is certainly the case that each individual environmental manager will face a different set of uncertainties – and, consequently, will tend to respond through their EM practices in different ways (see below). Here, one hypothetical example is used to illustrate this point. A poor farmer may face environmental uncertainty linked to ecological processes affecting the land. That uncertainty can also reflect over-exploitation of that land as the farmer attempts to maximize outputs. The maximization of outputs may be linked to production for the market and the need to increase production to compensate for falling commodity prices. Such market uncertainty is often made worse by political uncertainty. Poor farmers may be the victims of legal and coercive processes that, in aggregate, often leave them further marginalized in society. Further uncertainty may derive from broader sociocultural processes. Thus, changing consumption habits or tastes (as transmitted to the farmer through the market) may render increasingly difficult the choices that the farmer has to make about crop selection and production. The resulting EM practices of this environmental manager will be conditioned in complex ways by these diverse uncertainties.

This discussion has emphasized that all environmental managers are surrounded by uncertainty. Such uncertainty is not amenable to technological "quick fixes", and indeed may be inherent to multi-layered EM – a question considered throughout this book. Environmental managers respond to uncertainty using various means with the common goal of increasing predictability in EM.

Predictability

Environmental managers have sought to manipulate the environment in the pursuit of their livelihoods. Yet, as the preceding section highlighted, these efforts are usually dogged by different combinations of uncertainty. The response of environmental managers is to attempt to reduce uncertainty through the pursuit of predictability in EM. Chapter 7 explores the specific actions of different types of environmental managers as they articulate and seek to implement their environmental "policies". Here, it is important to specify the environmental, sociocultural, political and financial dimensions to the concept of predictability.

First, though, it should be stressed that predictability is not the same thing as "sustainability". A large literature examines the notion of sustainability (e.g. Slocombe 1993, Yin & Pierce 1993, Hilton 1994), and the associated concept of sustainable development (e.g. Barbier 1987, Redclift 1987, Lele 1991, Dovers & Handmer 1993, Reid 1995). This discussion does not bear repetition here. In this book, sustainability refers exclusively to the adoption of EM practices that do not lead to irreversible environmental degradation. Thus, the phrases of "sustainable EM" (cf. Turner 1988), and "unsustainable EM", are used throughout to differentiate practices that degrade the environment from those that do not.

Yet, although the notion of sustainability focuses on environmental considerations, the concept of predictability encompasses broader considerations of human–environment interaction that are linked to the pursuit of livelihood interests. The case of plantation forestry illustrates the difference between the two. When environmental managers create timber plantations, they are pursuing predictability in EM in that a process is initiated whereby income is derived from these plantations over a period of years. However, and as debate over plantation forestry illustrates (Sargent & Bass 1992, Wiersum 1995), it is as yet unclear whether these plantations are also environmentally sustainable. Evidence of soil erosion and nutrient depletion, for example, raises serious questions about the long-term ecological viability of this form of EM (Lohmann 1990). Thus, it can be seen how sustainability and predictability are not necessarily the same thing.

What then is the concept of predictability? At a general level, it is an attempt to minimize social and environmental uncertainty. Just as with the concept of uncertainty, predictability is not a single undifferentiated "good". Rather, it has several dimensions, and different environmental managers attach different significance to these dimensions. The latter point is taken up in the next section, whereas the environmental, sociocultural, political and financial dimensions of predictability are considered here.

Environmental predictability is sought by environmental managers to guarantee livelihood prospects based on carefully calculated use of environmental resources. Although there are many ways in which they may do so, three approaches are particularly important: the precautionary principle, and the sustained yield and carrying capacity approaches.

The precautionary principle is a response to uncertainty in the face of poorly understood environmental thresholds (O'Riordan & Cameron 1994).[1] The argument here is that exceeding these thresholds may lead not only to localized or regional catastrophes, but also to the elimination of human life on Earth. The precautionary principle is based on a paradox. To enhance environmental predictability in EM, there is a general need for "adequate" environmental data. Yet, such data are often unavailable or exist only for selected areas (Grant & Jickells 1995, Clayton 1995). Interpretation of environmental data can be a further problem (see, for example, Grainger 1993, and Brown & Pearce 1994, on tropical deforestation). The precautionary principle seeks to get around these difficulties by suggesting that limited data is still useful as an indicator, enabling the adoption of EM practices that safely fall beneath critical environmental thresholds. It suggests a strategy that recognizes that the benefits obtained through excessive environmental exploitation may be outweighed in the long term by the costs associated with exceeding environmental thresholds resulting in irreversible damage. These costs may include reduced livelihood opportunities – impoverished fishing communities adversely affected by overfishing, for example. They may also encompass problems related to ill health caused, for instance, by the build-up of toxins in drinking water. The realization of such costs may encourage environmental managers to err on the side of caution by limiting environmentally harmful practices. The case of European Union (EU) drinking water regulations, which were established on the basis of the precautionary principle, is illustrative in this context (Dieter 1992, Boehmer-Christiansen 1994b). Thresholds are set by state environmental managers at relatively "low" levels, believed by scientific advisors to be well within toxic thresholds that might adversely affect human health. The precautionary principle is, thus, a "conservative" approach that eschews maximum exploitation as part of the pursuit of environmental predictability.

In contrast, the sustained yield and carrying capacity approaches seek to enhance such predictability by pegging environmental use at a maximum level, yet still hoping not to degrade the environment. The principle of sustained yield is applied by environmental managers to specify the upper limits of resource extraction (taking from the environment) (Ingold et al. 1988, Maser 1994, Wiersum 1995, Xu et al. 1995). It has perhaps been applied most effectively in the promotion of environmental predictability by environmental managers operating at a relatively small scale. For example, it has been used in estimating commercial production levels in forest plantations, where it is reasonably straightforward to estimate the annual growth of timber and the ecological requirements of the plantation as a whole. Calculations are possible

---

1. Environmental thresholds are limits beyond which it is assumed that irreversible environmental degradation may ensue.

that may enable a maximum allowable cut of timber to be established in a given area, consistent with long-term forest maintenance. The timber harvest is set at that threshold level which, in theory, balances maximum exploitation with long-term conservation of the resource (Sargent & Bass 1992, Maser 1994, Wiersum 1995). This principle has not been successfully applied at anything other than the local level, because of the greater unreliability of environmental data at regional or global scales (Mitchell 1989, Rees 1990). Thus, the ability of the principle of sustained yield to specify environmental thresholds, as part of an attempt to promote environmental predictability, is largely scale-dependent (see, for example, Fairlie 1995, and Rettig 1995, on world fisheries).

Yet, even at the local level there are doubts about the utility of sustained yield calculations. What may, for example, be applicable in simplified forests such as plantations becomes problematic as the forest increases in complexity. In particular, the interaction of a multitude of species may influence timber growth in unpredictable ways (i.e. shading, competition for water and nutrients) (Whitmore 1990). The sustained yield principle, therefore, is of only limited use to environmental managers around the world.

The carrying capacity approach has also been more useful at the local level, where it is often used to estimate the maximum number of people or animals that can be supported by environmental resources in a given area. As with the principle of sustained yield, this approach is designed to enhance environmental predictability for specific environmental managers. For example, the carrying capacity approach has been applied to protected areas management by park managers as part of an attempt to regulate the impact of people on fragile environments (Stankey 1980, Hammit 1993, Kliskey 1994). However, this approach has been most extensively tested in the context of rangeland management (Abel & Blaikie 1989, Walker 1993). Here, it seeks to estimate the maximum number of grazing animals that can be stocked on a given area without causing irreversible environmental degradation. In central Somalia, for example, Baas (1993) established the maximum animal stocking rate on the basis of fodder availability, enabling a rough estimation of carrying capacity to be made. There are certainly considerable difficulties associated with this task (Homewood & Rodgers 1987, Scoones 1989, DeLeeuw & Tothill 1990). Carrying capacity calculations may be influenced by such factors as seasonal change, range condition (quality of the grass), annual climatic variations, droughts, previous use, and changes in human practices (Foran & Stafford Smith 1991, Mwaloyosi 1991, Warren 1995). As with the principle of sustained yield, there seem to be clear limits to the applicability of the principle of carrying capacity.

Thus, both the sustained yield and carrying capacity approaches aim to specify maximum resource exploitation levels beyond which environmental degradation is seen to ensue. In contrast, the precautionary principle errs on the cautious side concerning possible environmental thresholds. All of these approaches, nonetheless, seek to promote environmental predictability for environmental managers. Yet, as environmental managers pursue their policies and practices, they also aim to promote sociocultural, political and financial predictability. As Chapter 1 noted, EM is a multi-layered process in which environmental managers interact with the environment, but also with each other, in the pursuit of their livelihoods. For this reason, EM can never be fully understood without a consideration of the manner in which environmental managers interact with each other. Such interaction often leads to uncertainty, as noted above. It is also associated with the responses of environmental managers to uncertainty, which emphasize the promotion of sociocultural, political and financial predictability.

The pursuit of sociocultural predictability may have a direct bearing on the policies and practices of environmental managers. Such predictability is sought through a variety of social activities and networks that may encompass, for example, social clubs, marriage, kinship groups, religious meetings or voluntary labour on communal projects (Berry 1989). These activities, seemingly far-removed from the concerns of EM, may be nonetheless absolutely crucial to the long-term success of EM practices. In effect, these diverse social activities and networks serve to build a sense of goodwill and reciprocal obligation among selected environmental managers, which may be central to the protection of their livelihoods. For example, many farmers living in drought-prone areas have long been reliant in times of dearth on personal relationships with other environmental managers. During emergencies, farmers may be able to call upon other members of the local community to provide income, food or labour to assist them in their EM practices (Berry 1989, Watts 1983a).

Similarly, political predictability may be sought by environmental managers to ensure continuity in their EM practices. At a basic level, they usually engage in political processes in order to eliminate potential threats to their livelihoods, as well as to create new EM opportunities. Such activity can take many forms. It may, for example, encompass efforts by TNCs to persuade, cajole or bribe powerful political leaders or bureaucrats to minimize taxation and environmental regulation, as well as to provide funds for infrastructure development and other EM opportunities (Korten 1995, Moody 1996). Political predictability may also reflect the promotion by grassroots organizations (often representing poor environmental managers) of democratic conditions in society, as part of an attempt to combat local oppression and aid the EM policies and practices of these managers (*The Ecologist* 1993, Pretty 1995). Political predictability in the latter case may be reflected in secure land rights, the absence of state-sponsored terrorism, the elimination of exploitative local elite demands, and local control of community environmental resources.

Financial predictability may be another means by which environmental managers seek to ensure the continuation of their livelihoods. This may be associated with the quest to maintain a stable income from EM activities. In the case of poor farmers, for example, a classic response to fluctuations in market prices has been the adoption of a diversified cropping regime. Rather than planting only one or two crops, these environmental managers may plant many different crops with the aim of ensuring that a continuous and stable livelihood is attained (Pretty 1995). This quest for financial predictability is concerned not so much with maximizing income opportunities through EM practices as it is ensuring that basic livelihood needs are met (cf. Scott 1976). Other environmental managers may seek to maximize income as the central plank in their quest for financial predictability. The best example here relates to TNCs, which usually structure their EM policies and practices with the overriding objective of profit-maximization (Korten 1995, Welford 1996). For TNCs, financial predictability is all about increasing wealth so as to enhance future EM opportunities. It may, of course, be the case that this strategy results in pervasive environmental degradation, thereby reiterating the earlier point that predictability is not necessarily the same as "sustainability".

The preceding discussion has highlighted the environmental, sociocultural, political and financial predictability that many environmental managers may seek in the pursuit of their livelihoods. As discussed below, they often follow different routes in the pursuit of predictability. Yet, what needs to be noted here is the fact that environmental managers may pursue simultaneously the different dimensions to predictability, although not all environmental managers pursue all types of predictability concurrently. Again, a hypothetical example may be useful to illustrate the general point. A wet-rice farmer may pursue predictability along multiple dimensions through the construction and maintenance of an intricate network of rice terraces and associated irrigation channels. Insofar as this network helps to prevent soil erosion and desiccation, the farmer seeks to increase environmental predictability. Irrigated terraces may also enhance financial predictability in as much as they facilitate stable production and, hence, potentially stable income for the farmer. Since these terraces require a considerable input of labour – both for initial construction and subsequent maintenance – they place a premium on continuous cooperation among farmers, usually necessitating the creation of complex reciprocal arrangements and communal institutional mechanisms (Ostrom 1990). These arrangements usually enhance sociocultural predictability for the farmer through the generation of local goodwill and reciprocity. Finally, irrigated terraces are often associated with localized control over scarce environmental resources through irrigation committees in which the farmer may have an important say. To the extent that these local committees are accountable to the farmer and possess the requisite power to control water access, political predictability is also thereby enhanced.

This section has emphasized the different dimensions to the concept of predictability. All environmental managers pursue predictability through their policies and practices, but such endeavours are conditioned by diverse considerations that may differ between types of environmental managers. The following section takes a closer look at this issue by exploring some of the ways in which environmental managers respond to uncertainty through a variety of efforts to promote predictability.

Uncertainty, predictability and environmental managers

So far, this chapter has sought to explain the concepts of uncertainty and predictability, in order to suggest their utility in understanding the operations of multi-layered EM. The objective of this section is to explore the possible meaning and significance of these concepts for different types of environmental managers.

Table 2.2 illustrates generally how uncertainty and predictability may be related to the different types of environmental managers that are the focus of this book (see Ch. 1). In the table, environmental, sociocultural, political and market uncertainty and predictability are linked to the state, environmental NGOs, farmers, TNCs, hunter–gatherers and international financial institutions in terms of low-, medium- and high-intensity relationships. What follows is a selective discussion of the differing relationships summarized in Table 2.2.

As noted in Chapter 1, the state is often seen as being an important environmental manager, whose social and environmental activities can have a considerable impact on the activities of other environmental managers. As Table 2.2 illustrates, the state's comprehensive EM role leaves this actor highly exposed to most of the different types of uncertainty discussed in this chapter. As would be expected, political uncertainty is a prime concern for state environmental managers. Such uncertainty may take the form of programmatic and personnel changes arising from the electoral process, notably in democratic countries. It may also be linked to violence directed against the state and its representatives, which might culminate in the overthrow of the existing political regime. These possible changes often have important ramifications for state EM as new ideologies, interests and personnel seek to shape the direction of certain EM practices (Peluso 1992). Less self-evident, is the high degree of sociocultural uncertainty that states usually face. Especially under democratic political regimes, the fate of state leaders often depends on their ability to win and retain public trust. The failure to do so often leads to political defeat, and potentially altered EM practices.

States, nonetheless, have a variety of ways in which they seek to increase predictability in EM. State leaders seek to enhance political predictability through efforts designed to ensure electoral success. With potentially substantial financial and human resources at their disposal, they seek to appeal to voters through EM policies and practices designed to satisfy the interests and concerns of the public (Lowe & Goyder 1983). States may also seek environmental predictability through measures, for example, designed to prevent catastrophes or halt degradation. Typically, however, state leaders emphasize immediate concerns linked to, say, political predictability over "long-term" factors associated with environmental predictability.

Environmental NGOs provide an interesting comparison to states in terms of their relationship to both uncertainty and predictability. They have developed as a societal response to growing public concern about environmental

Table 2.2 Uncertainty, predictability and environmental managers.

Environmental managers	Uncertainty				Predictability			
	Environmental	Sociocultural	Political	Market	Environmental	Sociocultural	Political	Financial
State	MEDIUM Global climate change	HIGH Public trust in the state	HIGH Elections	HIGH International trade flows	MEDIUM Prevention of environmental catastrophes	MEDIUM Environmental education	HIGH Electoral success	HIGH Increasing gross national product
Environmental NGOs	LOW Indirect interaction with environment	MEDIUM Attitudes of members	HIGH Flexibility of state regulations	LOW Indirect link to market	HIGH *Raison d'être* for environmental NGOs	HIGH Changing environmental attitudes	HIGH Lobbying	MEDIUM Increasing importance of finance in environmental campaigns
TNCs	MEDIUM Environmental resource base	MEDIUM Consumer confidence in products	MEDIUM State environmental regulations	HIGH Price of environmental resources	MEDIUM Timber plantations	MEDIUM Manipulating consumer behaviour	MEDIUM Lobbying	HIGH Profit
International financial institutions	MEDIUM Environmental degradation in funded projects	MEDIUM Lack of support of projects by local community	HIGH Lack of state support for projects	MEDIUM Fluctuating prices for project-based commodities	MEDIUM Prevention of environmental degradation through project-based support	MEDIUM Support of projects by local community	HIGH State support for projects	MEDIUM Profit
Farmers	HIGH Drought	MEDIUM Lack of neighbour or community support at times of hardship	MEDIUM Changing state environmental regulations	HIGH Commodity prices	HIGH Prevention of environmental degradation	MEDIUM Neighbour or community support	MEDIUM Seek to influence state regulations	HIGH Stable/increased income
Hunter-gatherers	HIGH Fluctuating food resources	LOW Relative cultural homogeneity	HIGH Increasing marginalization	LOW Weak market integration	HIGH Ensure survival of family/group	HIGH Group support	MEDIUM Ancestral domain claims	LOW Weak market integration

problems. As Chapter 1 discussed, their role in EM is indirect: they seek to alter the policies and practices of other environmental managers. As with the other types of environmental managers discussed in this book, they are not immune to the effects of different types of uncertainty. As with states, environmental NGOs must necessarily confront great political uncertainty, usually stemming from the rules and regulations that the state imposes to regulate their activities in society (Wapner 1995). In contrast, and unlike many other environmental managers (e.g. TNCs or farmers), environmental NGOs face relatively little market uncertainty, given that they are not primarily preoccupied with the business of buying or selling commodities, and hence are usually less exposed to market fluctuations than other environmental managers are.

Environmental NGOs perhaps place the greatest emphasis on promoting environmental and sociocultural predictability. On the one hand, the quest for environmental predictability – understood by environmental NGOs as sustainable EM – is their very *raison d'être*. On the other hand, environmental NGOs believe that the transformation of public environmental attitudes is crucial to the attainment of sustainable EM, thereby also highlighting this actor's emphasis on sociocultural predictability (Laferriere 1994). This pursuit of predictability is reflected in a concern to promote the long-term wellbeing of the Earth and associated campaigns to alter existing patterns of human–environment interaction.

TNCs are usually seen as being the antithesis of all things that environmental NGOs believe in and fight for. These corporations have assumed growing importance as environmental managers in the measure that their economic power has increased. The considerable power of TNCs does not make them immune to social and environmental uncertainty (Pearson 1987). For example, TNCs are quite vulnerable to market uncertainty. This may affect the use of environmental resources (e.g. by making it less worthwhile to exploit resources whose value has fallen on the world market) and may also influence the marketing of end-products (e.g. consumer demand for "greener" products). TNCs are also subject to political uncertainty insofar as their operations are regulated by states. Consequently, they constantly face the prospect of external interference with their profit-maximizing activities. The considerable economic power and potential mobility of TNCs nonetheless enable them to mitigate state regulations in their EM practices, thereby somewhat attenuating political uncertainty.

It is TNCs that are most closely associated with EM practices that seek to promote financial predictability understood almost entirely by them in terms of profit maximization. Indeed, the *raison d'être* of TNCs is that of profit-maximization in a global capitalist system (Korten 1995). They nonetheless pursue predictability in other ways. TNCs increasingly pursue environmental predictability through EM practices designed to ensure a regular supply of environmental resources (e.g. through forest plantations). Yet, this pursuit should not be exaggerated in terms of resource conservation. Most TNCs pursue environmental predictability only to alleviate intensifying resource scarcity, which, in turn, may reduce profits.

As with TNCs, international financial institutions have had a major environmental impact as a result of their role as worldwide promoters of economic development. They are similar to environmental NGOs in that their EM role is largely indirect. The EM role of international financial institutions is derived from the selective provision of financial assistance to other environmental managers such as states or farmers. International financial institutions have developed over the past 50 years at the behest of the USA and other leading Western states (Rich 1994). Not surprisingly, therefore, international financial institutions face considerable political uncertainty associated with the need to continually solicit the financial backing of these powerful states for their EM projects. In a context of growing worldwide environmental degradation linked to the development process, international financial institutions have needed also to confront growing environmental uncertainty. Paradoxically, this uncertainty is often linked to the misguided projects that they have supported over the years (e.g. many Green Revolution projects). Such uncertainty may not be as strong for international financial institutions as it is for other types of environmental managers such as poor farmers or hunter–gatherers, who bear the brunt of many environmental problems directly. Nevertheless, environmental uncertainty is exerting an increasingly important influence on the EM activities of international financial institutions.

International financial institutions have sought predictability in their EM practices traditionally through state sponsorship. Such political predictability is critical to the largely indirect efforts of international financial institutions to effect changes in the policies and practices of other environmental managers. Yet, increasing environmental uncertainty surrounding projects sponsored by international financial institutions has led to growing pressure on these environmental managers to take into account local public concerns and interests. As such, the quest for sociocultural predictability has strengthened as international financial institutions seek the support of local communities where funded projects are implemented (World Bank 1992).

The growing attention paid by international financial institutions and states to local community concerns has emphasized the central importance of farmers (as well as fishers, nomadic pastoralists or shifting cultivators) in multilayered EM. Unlike environmental NGOs and international financial institutions, farmers have an immediate impact on the environment through their EM policies and practices orientated around the production of food and fibre. Farmers are more vulnerable than most other environmental managers to uncertainty derived from environmental and market exposure (Pretty 1995). Environmental uncertainty may be linked to the variability of the natural environment (e.g. droughts), but also to environmental degradation associated with over-exploitation of environmental resources. Market uncertainty may compound the plight of farmers in that it often links their livelihoods directly to price fluctuations that are usually beyond their control.

For these reasons, financial predictability is habitually a top priority of farmers. Although some farmers seek to maximize profits through large-scale monocultural production techniques, others use smaller-scale polycultural EM strategies to ensure that basic livelihood needs are always met. Yet, the typically high exposure of farmers to environmental uncertainty often leads them also to pursue sociocultural predictability through local social institutions and networks, with the aim of obtaining community support in their EM practices (e.g. terracing, maintenance work) at times of hardship (Chambers 1987).

As with farmers, hunter–gatherers are a type of environmental manager faced with great environmental uncertainty. They also interact directly with the environment in their livelihood pursuits. Indeed, more than any other type of environmental manager, they are most closely associated with "the environment" in the popular imagination (Denslow & Padoch 1988). That close association is reflected in the need for hunter–gatherers continually to adapt their EM policies and practices to shifting environmental conditions, for otherwise hardship, if not death, may result. Yet, another typical feature of hunter–gatherers – their relative lack of integration in the global capitalist system – means that they face low market uncertainty; for example, they generally do not produce goods for the market and are usually not affected by price fluctuations.

Not surprisingly, hunter–gatherers usually place a premium on environmental predictability. Highly complex EM policies are devised in order to ensure that basic livelihood needs are met from the environment, the vicissitudes of natural environmental processes notwithstanding. Yet, the ability of hunter–gatherers to pursue their livelihood is compromised increasingly by the policies and practices of states who claim jurisdiction over their territories (Hong 1987). As such, hunter–gatherers are placing increasing emphasis on the pursuit of political predictability, notably through the articulation of "ancestral domain claims" that directly counter the claims of states.

Table 2.2 provides a general scheme only for understanding some of the possible interactions between uncertainty, predictability and environmental managers operating in multi-layered EM. As such, it does not explore the potential internal complexity of different types of actors (e.g. different agencies, political leaders within state EM), and how such complexity is related, in turn, to questions of uncertainty and predictability. Although this book addresses aspects of this subject, its focus on developing a more general understanding of the dynamics of multi-layered EM obviates the need for detailed discussion of how individual environmental managers respond to uncertainty.

The preceding discussion has sought to selectively highlight the relationship between uncertainty, predictability and the different types of environmental managers considered in this book. Subsequent chapters will elaborate this general relationship. Chapter 7, in particular, will be concerned to take up questions raised here surrounding the pursuit of predictability in EM through a detailed discussion of the EM policies of environmental managers.

Conclusion

This chapter has explained the concepts of uncertainty and predictability in order to clarify the context within which environmental managers operate. The environmental, sociocultural, political and market uncertainties facing environmental managers were described. This discussion has highlighted not only the differentiated nature of uncertainty, but also that different types of environmental managers may face different combinations of uncertainty. The response has been the pursuit of predictability in terms of environmental, sociocultural, political and financial dimensions. Here, too, the differentiated nature of the quest for predictability was emphasized.

Part I of this book has set out the analytical framework that informs the argument of this book. Chapter 1 established the idea of EM as a multi-layered process in which state and non-state environmental managers interact in the course of pursuing their EM objectives. Chapter 2 then suggested that those objectives revolve around the pursuit of predictability in a context of social and environmental uncertainty. The remainder of this book uses this framework to understand the process of EM and the ways in which environmental managers seek to pursue different policies and practices in keeping with their livelihood interests. Part II explores the relationship between, on the one hand, intensifying human use of the environment and the environmental attitudes and world-views of environmental managers and, on the other, increasing social and environmental uncertainty in EM. In contrast, Part III examines how political, market and policy processes have a bearing on the pursuit of predictability by environmental managers operating in multi-layered EM.

PART II

INCREASING UNCERTAINTY

Chapter 2 showed that state and non-state environmental managers similarly face social and environmental uncertainty. Their response has been to pursue predictability in human–environment interaction. Yet, the history of this interaction has been one in which exploitation often has prevailed over conservation, with environmental degradation as the frequent result. Indeed, today environmental problems have assumed global proportions, reflecting a legacy of unsustainable practices in multi-layered EM.

To understand why this situation has developed, it is necessary to explore the differing EM practices and attitudes that are at the heart of environmental degradation. To this end, Chapter 3 examines how growing human numbers and intensifying EM practices have contributed to cumulative environmental degradation that poses a particularly great challenge to contemporary environmental managers. Chapter 4 emphasizes those environmental attitudes, worldviews and discourses that might be at the base of such EM practices. The argument is that past and present EM practices and attitudes have contributed to greater overall uncertainty in EM.

CHAPTER 3

Human–environment interaction

Humankind must use the environment in order to survive. However, without conservation practices, long-term environmental degradation sets in, threatening human survival. This enduring tension between exploitation and conservation has always been at the heart of EM. Yet, a central characteristic of human progress has been the ever more intensive use of the environment, leading to increased uncertainty. The discussion in this chapter is designed to provide an overview of this process, so as to set in context the analysis of different types of environmental managers operating in multi-layered EM contained in subsequent chapters.

There are two aspects to the historical development of human–environment interaction. First, a growing human population has had a generally greater impact overall on the environment. Secondly, changes in life-styles associated with technological innovation have resulted in a greater per capita human environmental impact. These historical changes have had a major impact on how EM is practised. As the number of human beings has increased, so too has the number of environmental managers. The growing complexity of human societies, and especially the rise of the state as a leading environmental manager, has been reflected in the growth of state EM. Indeed, such has been its growth that some scholars have equated EM solely with state EM (see Ch. 1). If, as this book argues, this is not the case, it is nevertheless true that EM as a process has become increasingly complex and multi-layered over time. Indeed, it is a central paradox of EM that intensifying human use of the environment has served to enhance uncertainty, thereby undercutting the efforts of environmental managers to promote predictability. Owing to growing complexity in human–environment interaction, the ability of environmental managers consistently to attain predictability in their EM policies and practices is becoming increasingly remote.

Population and environmental management

The growth of the human population has simultaneously intensified environmental use and increased the number of environmental managers. This trend has enhanced uncertainty as more environmental managers interact with (and often degrade) the environment. Yet, the relationship between population and intensified environmental use is far from straightforward. We now review the debates surrounding this issue as part of a clarification of the significance of the "population question" for multi-layered EM.

The third International Conference on Population and Development, held in Cairo in September 1994, highlighted some of the tensions surrounding the link between population growth and environmental degradation. This conference, which brought together representatives from 176 states and almost 4000 environmental NGOs, highlighted the seriousness of unchecked population growth, with a world population of 6 billion increasing by about 100 million people a year (Johnson 1994). Despite such evidence, delegates from the Vatican and certain Islamic countries denied the need for drastic measures to curtail birth rates. However, a general consensus developed that current population growth is jeopardizing humankind's ability to manage the environment sustainably. The conference reiterated the growing uncertainty that characterizes human–environment interaction and that limits the ability of environmental managers to enhance predictability in EM.

There are two issues at the heart of current debates about population growth. First, there is the question of whether increasing human numbers necessarily result in environmental degradation, or whether increases may even form a prerequisite for sustainable EM. Secondly, there is the issue of whether human consumption levels, as opposed to human numbers, are ultimately to blame for environmental degradation.

Rising human numbers may lead to environmental degradation, but what is causing most concern is the rate of increase (Brown & Kane 1994, Lutz 1994). It took millennia for humanity to reach 1 billion people by 1800, whereas it took only a further 130 years to add a second billion. Yet, the third billion was attained in only 30 years, and world population in the late 1990s has exceeded 6 billion. This increase has occurred in a context of more or less constant life-supporting systems of Earth. Although the latter change over long timescales, such change is negligible in human time-frames.

Pessimism about population growth is typically associated with Thomas Malthus (Lowe & Bowlby 1992). In 1798, he argued that human population would inevitably increase faster than agricultural production. Malthus argued that, whereas human populations increased exponentially, the quantity of natural resources was fixed. Humanity's future was one of increasing misery and catastrophic famines, unless population growth was controlled. Rapid post-war population growth – especially in economically less developed countries (ELDCs) – combined with perceptions of growing ecological crises in economically more developed countries (EMDCs), led to the growing popularity of the Malthusian viewpoint in the 1960s and 1970s (e.g. Carson 1962). Writers such as Ehrlich (1970) and Heilbroner (1974) warned of a "population bomb" leading to inevitable catastrophe. This "neo-Malthusian" argument was seemingly lent scientific respectability through the work of Meadows et al. (1972) and

Barney (1980). Using the latest computer technology, this research specified "limits to growth" based on projections of population growth and resource depletion. Through quantitative means, neo-Malthusians sought to reiterate Malthus's original predictions.

The neo-Malthusian argument was condemned by critics from the political Left and Right. The argument on the Right was that this approach was conceptually flawed and unnecessarily pessimistic about human ingenuity and adaptability (Beckerman 1974). Simon (1981), for example, viewed increasing human numbers as an opportunity rather than a constraint: more people means a greater human capacity to innovate and implement new ideas (notably in regard to managing the environment), and not simply more mouths to feed (see also Holloway 1995). In contrast, left-wing writers criticized the authoritarian political implications of the need to enforce population regulations to protect the environment (Buchanan 1973). Enzensberger (1974) and Harvey (1974) highlighted that population control was based on draconian state intervention in even the most intimate aspects of human life. To achieve such intervention would require a global Leviathan with powers on a scale as yet not even remotely obtained, and would subordinate individual choice to a collective will in the management of the environment.

Approaching the issue from a different direction, other researchers have turned the neo-Malthusian argument on its head. Boserup (1993) argued that population increase often can be accommodated by environmental managers (i.e. farmers) through intensified agrarian practices. She argued that "in many cases the output from a given area of land responds far more generously to an additional input of labour than assumed by neo-Malthusian authors" (ibid.: 14). This argument focuses on the relationship between population growth and food production. Tiffen et al. (1994) have further argued that population increase, rather than being a cause of environmental degradation may, in fact, promote sustainable EM. Based on a Kenyan case study, they make the point that certain EM techniques, such as terracing, are based on "economies of scale". Only when population densities reach a certain threshold can large investments of labour and material for sustainable EM become feasible.

Boserup (1993) and Tiffen et al. (1994) are useful correctives to the neo-Malthusian perspective, and have intensified the debate on the EM implications of population increases (Harrison 1993). However, the Tiffen et al. (1994) study is based on only one region, and one specific issue of environmental degradation. The general applicability of their research has yet to be proven. Further, this study does not tackle the broader issue of increases in the total world population. A problem of scale may be encountered: what seems to apply at a local level may not be relevant at regional and global levels. There is yet the issue of increased populations having a greater total adverse environmental impact than smaller populations, when all factors are taken into account. Even if it can be shown, for example, that larger populations do not necessarily lead to overuse of resources (i.e. taking from the environment), there is still the issue of increased pollution levels associated with larger populations (i.e. adding to the environment).

It must be emphasized that there are also examples where relatively small populations have led to large-scale environmental degradation (Blaikie & Brookfield 1987). In New Zealand, for example, between the tenth and nineteenth centuries small numbers of Polynesian settlers destroyed large tracts of forest and caused large-scale extinction of species. Utilizing fire as a technique to flush out game and for agricultural purposes, these Polynesians, who numbered no more than 300000 people at any one time, cleared 60000 km^2 of native forest on a quarter of the country's land surface (Cumberland 1961, McGlone 1983). This shows that the relationship between population size and the scale of environmental degradation is not necessarily evident.

Considering the link between population, environmental degradation and EM also means accounting for per capita consumption levels. In this regard, it is useful to distinguish between "people overpopulation" and "consumption overpopulation" (Ehrlich & Ehrlich 1990). Whereas the former refers to total population numbers, the latter directs attention to unequal human consumption of the environment (taking from and adding to the environment). Specifically, it is concerned with the difference between needs and wants. Needs ensure basic survival (food, shelter), whereas wants go beyond these basic levels (television, car). Consumption overpopulation, therefore, highlights per capita rather than aggregate use of the environment.

There are indeed sharp discrepancies in this area. For example, the 23 per cent of the world's population living in the 33 EMDCs use 80 per cent of the world's mineral and energy resources (World Resources Institute 1995). The USA has only 4.8 per cent of the world's population, but uses about one third of the world's non-renewable energy and mineral resources, and generates one third of the world's pollution. The USA also uses 370 times more energy per capita than does Sri Lanka, and the 24 million new Americans born between 1984 and 1994 will utilize more energy and resources than the entire population of Africa. On an annual basis, the car-based commute into and out of New York alone uses more oil than the whole of Africa, excluding South Africa (Edge & Tovey 1995). In aggregate, these data highlight the enormous disparity in consumption between EMDCs and ELDCs.

Discrepancies in individual consumption levels within countries are similarly, if not more, unequal. Throughout the world the economically affluent contribute disproportionately to environmental degradation through conspicuous overconsumption. Whereas the majority of ELDC populations are rarely able to meet much more than basic livelihood requirements, their affluent counterparts often indulge in consumer-driven Western life-styles that are at the centre of much environmental degradation. This complicates further the prospect for predictability in multi-layered EM.

Consumption overpopulation highlights the important point that environmental degradation is not necessarily linked to population growth. Yet, this debate also has a direct bearing on EM in several respects. First, it highlights the uncertainty surrounding the EM implications of population growth. If population growth is the prime source of environmental degradation, then environmental managers must find ways to stem such growth. In contrast, if human consumption per capita is the main culprit, then environmental managers need to address the implications of unequal use of the environment.

Secondly, and irrespective of which factor is the more important, the population–consumption debate highlights the increasing uncertainty associated with human–environment interaction. Whereas the specific weight that should be attached to each factor is still contested, what is incontestable is that human use of the environment has intensified dramatically in recent centuries, thereby increasing the difficulties that many environmental managers face in the context of growing uncertainty.

Thirdly, the debate has raised questions about which environmental managers are empowered and which are not with respect to regulating human numbers and consumption. Despite differing viewpoints, many scholars assume that the state must play a central role in resolving theses problems (cf. Corbridge 1986). States are seen to be the leading environmental manager in enforcing population control policies or, alternatively, are called upon to limit consumption. Indeed, states have already sought to control population increase (e.g. China, India). However, the state does not always take the lead on this issue. There are many examples where non-state environmental managers have carefully regulated population numbers as part of the quest to reduce uncertainty in EM. At one extreme, for example, some indigenous people of northern Greenland and Canada have committed female infanticide in order to keep their population size within environmental limits (Miller 1994).

The population debate has important implications for EM. Yet, as the discussion of excessive consumption highlights, it is not simply human numbers that are at issue here, but also levels of consumption linked to intensifying human use of the environment over time.

Intensified environmental use

The ways in which intensifying use of the environment relates to environmental degradation are best illustrated with reference to certain types of human–environment interaction: huntering–gathering, agriculture and industrial production. These different types of interaction involve different forms of environmental use. They highlight how differing types of human–environment interaction can create different environmental impacts and, in turn, are associated with different EM policies and practices. Further, they illustrate how environmental managers have responded to, but often have also increased, environmental uncertainty in the context of multi-layered EM.

The nature and scope of uncertainties faced by environmental managers operating under these different forms of resource extraction and environmental use have been different. Hunter–gatherers and agriculturists have been confronted traditionally by uncertainties associated with local and regional social and ecological issues relating to resource scarcity and conflict. For example, agricultural societies have faced local or regional environmental uncertainty in the form of droughts or crop invasions by pests, and sociocultural uncertainty has often arisen out of conflicts over land rights (see Ch. 2). In contrast, industrial societies generally confront uncertainties associated with resource scarcity, but also large-scale, if not global, environmental degradation. As the world becomes increasingly industrialized, the adverse environmental effects of intensifying human use of the environment have become all too apparent. Although all three types of human–environment interaction persist today, it is the industrial basis of the global capitalist system that is of greatest contemporary significance for environmental managers.

Hunting–gathering

Contemporary environmental degradation has prompted the search for alternative EM policies and practices. This search has emphasized the possibilities for sustainable EM in traditional hunter–gatherer practices. Rather than merely an early form of human–environment interaction that persists at the periphery of modern societies, hunting–gathering has been touted as a potential model for a sustainable EM of the future (see Beinart & Coates 1995).

The environmental impact of hunting–gathering is the subject of considerable debate, with implications for the understanding of hunter–gatherers as environmental managers. Some argue that hunter–gatherers have struck the right balance between exploitation and conservation, and may be seen as the "original" sustainable environmental managers (Devall & Sessions 1985). As popularized by Rousseau in the eighteenth century, this notion has been enshrined in the ideal of the "noble savage" – that is the idea that hunter–gatherers live in harmony with their environment in a way that "civilized" societies have long since lost. The ideal of the noble savage has persisted in EMDCs to the present, often underpinning how many "traditional" non-European societies have been portrayed by EMDC writers (Putz & Holbrook 1988).

This view has been criticized by those who argue that hunter–gatherers often failed to practise sustainable EM. These critics allege that the EM practices of hunter–gatherers often resulted in environmental degradation and social hardship (Sauer 1956, Simmons 1989). Indeed, and contrary to the traditional perspective, recent research has shown that hunter–gatherers with low population numbers could have significant adverse environmental impacts (Cronon 1983, Edwards 1988, Flannery 1990).

The example of Aboriginal hunter–gatherers in Australia sheds light on the debate surrounding the sustainability of EM practices of hunter–gatherers. Nowhere else in the world have hunting-gathering societies persisted for as long as 50000 years in virtual isolation from other societies (Head 1993). Further, natural changes to the Australian environment during this period have been relatively minor, enabling scientists to obtain a clearer picture of human–environment interactions. Therefore, the Australian situation allows us to evaluate competing claims regarding sustainable versus unsustainable EM practices of hunter–gatherers.

Australian Aborigines often are considered to be a prime example of sustainable EM (Jones 1991). The survival of the hunter–gatherer life-style for over 50000 years in the harsh Australian environment suggests that the pursuit of predictability by these environmental managers resulted in sustainable EM. The argument here is that such self-conscious EM practices were embedded in Aboriginal culture. The notion of the "dreamtime" – which specifies how the world was created as a manifestation of spirituality and which locates human existence within this spiritually conceived environment – can be construed as an attempt, in part, to ensure the protection of some elements of the landscape from over-exploitation. Places were embedded in the dreamtime, such as waterholes or forest groves, where the hunting of animals was constrained. Aboriginals also believed that humans were part of this spiritualized landscape. Land could be used, but never "owned" (Young 1992). As a result, EM was not based on practices linked to ownership (a central concern of environmental managers in capitalist societies), but on a spiritually based system of communal resource use. The EM policies of Australian Aborigines reflected these priorities (see also Ch. 7).

Other research has suggested that these environmental managers may have also been capable of degrading their environment. Pollen and charcoal analyses point to the destruction of forests by fire (Horton 1982, Kershaw 1986, Head 1989, 1993). Aboriginal burning is held responsible for this destruction, particularly in areas where natural fires were rare. Further, Aborigines may have played a role in the extinction of large mammals, birds and reptiles (Gillespie et al. 1978, Wright 1986, Flannery 1990). Indeed, evidence from New Zealand and Europe corroborates these claims of a "prehistoric overkill" by hunter–gatherers (Cumberland 1961, McGlone 1983, Martin & Kline 1984).

Thus, the practices of hunter–gatherers may not necessarily be synonymous with sustainable management of the environment. Indeed, examples from tropical forests reinforce this point. Ever since European explorers first encountered hunter–gatherers in tropical forests, the conventional wisdom has been that these environmental managers have adapted to, rather than modified, their environments. Yet, as with the Australian Aborigines, recent research has suggested a more active EM role for forest-dwelling hunter–gatherers. Whether it be the Penan in Malaysia or the Kayapo in Brazil, it is suggested that these environmental managers have skilfully manipulated their forest environments, but in such a way that it would not deplete future supplies (Hong 1987, Hecht & Cockburn 1989). Such "low-impact" EM practices seek to protect forests as the basis of a sustainable livelihood for increased predictability in EM.

These low-impact EM practices may be deceptive insofar as they conceal extensive manipulation of the forest. Hunter–gatherers have often manipulated species composition by encouraging the growth of useful species at the expense of others. Indeed, the impact of hunter–gatherers may have been so extensive historically that there is hardly any "unmanaged" tropical forest left, even in such vast areas as Amazonia (Meggars 1971). Hunter–gatherers may have systematically modified species composition while maintaining overall forest cover largely intact.

Hunter–gathering practices illustrate one way in which humans have sought to reduce environmental uncertainty in EM. It can be argued that hunter–gatherers are more than passive bystanders in environmental change. It has also been emphasized that, despite Western stereotypes, there is no essential link between hunter–gathering as an EM practice and sustainable EM. Further, the examples highlight that, even with low population densities, environmental degradation may yet result – showing that even "low-impact" EM may be associated with environmental degradation.

The tendency to romanticize hunter–gatherers through the "noble savage" ideal has obscured the ways in which these environmental managers have self-consciously and actively pursued strategies designed to increase predictability in the face of a harsh and often highly uncertain environment. In this regard, the manipulation of tropical forest species, and even the strong association of landscape with "spirits", may be seen as attempts to devise environmental policies to balance exploitation and conservation such that the uncertainty that jeopardizes human survival is minimized (Olofson 1995). Although this quest for predictability in EM was not always successful, it nevertheless has formed the basis of one set of EM practices that even today is upheld in some quarters as the basis for sustainable EM. Yet, although certain practices are worth considering in contemporary contexts (i.e. non-clearfelling, selective logging), it is wrong to see hunter–gathering EM practices as a general blueprint for the future in a world that is largely shaped by other, much more intensive, EM practices, especially permanent agriculture and industrial production, which are blamed for much of today's environmental degradation.

Agriculture

In contrast to hunting–gathering, agriculture has had a more conspicuous impact on the environment. The clearance of millions of hectares of forest around the world by farmers to make way for permanent agriculture has been a central feature in EM over the centuries. At issue here is not so much whether an impact has occurred or not; rather, it is the severity and meaning of that impact that is at stake. However, the transformation of forest into field does not necessarily entail environmental degradation, but does represent the transition from one set of environmental conditions to another. When farmers fell forest, for example, they interrupt natural ecological processes and reduce biological diversity in order to maximize food output. Farmers, therefore, need to manage a much more fragile "artificial" environment than do hunter–gatherers (Harris 1978). This has considerable benefits in terms of additional food production as compared to reliance on gathering.

At the heart of this process is the farmer. This environmental manager is responsible for EM practices that encompass large areas of the Earth's land surface. Through these EM practices, farmers have sought to balance the need to produce food and fibre on the one hand, with the need to protect the environmental bases of such production on the other. Farmers have faced uncertainty just as much as hunter–gatherers have, yet the specific nature of such uncertainty has differed. For farmers, uncertainty has been linked to the fact that their livelihoods are derived from the transformation and intensive management of relatively small areas of land. This dependency on the land has resulted in two distinct types of environmental uncertainty in particular (see Table 2.2 in Ch. 2). First, uncertainty has derived from variability in the natural environment, especially in marginal areas (i.e. droughts, floods). Unlike hunter–gatherers who are relatively mobile and have more flexibility in seeking new means of sustenance, most farmers tend to be less mobile and less able to respond rapidly to sudden changes. Secondly, and especially in recent years, uncertainty has resulted from the ways in which farmers have sought to maximize production, notably through the use of fertilizers and pesticides. In the quest to reduce uncertainty over the ability to produce food, farmers have inadvertently generated new environmental uncertainties associated with the pollution of the environment (Carson 1962, Briggs & Courtney 1989).

Farmers have generally adopted two distinct EM strategies. The first has been favoured in areas with low population densities and limited productive capabilities, and is based on a system of rotational use. In many forested areas, farmers have pursued an EM strategy of shifting cultivation in which temporary plots (typically up to five years) are cleared in the forest. Because soil fertility usually declines rapidly, these farmers are forced to move periodically to new forested areas in order to allow the used area to regenerate itself. At some future date, once nutrients in the soil have been restored through natural process-es, cultivators might return to the same area to repeat the cycle. Although this EM strategy is variable in different cultural settings, the general principle is one in which agriculture takes place over an extensive area in order to avoid environmental degradation. Indeed, provided that there is enough land for cultivation and that adequate time is left for regeneration, this strategy is often seen as a prime example of sustainable EM (Dove 1983).

An equivalent EM strategy in semi-arid environments is that of nomadic pastoralism, where herders manage their environment through the rotational and periodic use of grazing areas by domestic animals. Just as shifting cultivators reduce the likelihood of environmental degradation by rotating plots, so nomadic herders aim to ensure that the range resource is managed sustainably through the regular movement of herds (Homewood & Rogers 1987, Baas 1993).

Whether it be shifting cultivation in forests or nomadic pastoralism in semi-arid areas, this EM strategy is a carefully calculated response to high environmental uncertainty in the local environment. Unlike hunter–gathering, this EM strategy is focused on the management of the environment for a limited range of agricultural products. Unlike permanent agriculture, rotational or periodic use of the environment by shifting cultivators and herders enhance predictability through the spatial shifting of production (Blackburn & Anderson 1993).

Under appropriate conditions, this EM strategy is sustainable, provided that enough time is allowed between initial exploitation and subsequent re-use. As a result of other environmental managers impinging on the resource base, this essential fallow period has been eroded in many areas (Hong 1987, Horowitz & Little 1987). For example, shifting cultivation in many parts of Southeast Asia has become untenable in the face of widespread land clearance for permanent agriculture, logging, mining and dam construction (Brookfield 1988, Bryant 1994a). The rotational EM practices described above may be doomed in a world of intensifying and permanent land uses.

In densely populated areas, farmers have pursued permanent agriculture. In the process, EM has been about the ever-more intensive exploitation of a strictly delimited area. Rather than seeking predictability through the EM practice of shifting cultivation, farmers involved in permanent agriculture rely increasingly on systematic and widespread use of such external inputs as fertilizers as a substitute for fallow periods. In the process, they have been able to increase production from the land dramatically, but in many cases only at the expense of increased uncertainty through environmental degradation (Knickel 1990, Robinson 1991). The fact that permanent agriculture is the most important land use on Earth means that these farmers are one of the most important types of environmental managers. The decisions of individual farmers over such issues as chemical use and land clearance may not have a major impact, but the aggregate impact of all these decisions may have severe environmental repercussions (Briggs & Courtney 1989, Rozanov et al. 1990, Rigg 1991).

At the heart of the success of permanent agriculture in expanding food and fibre production is the intensification of production. As a result, agricultural yields have increased dramatically. The environmental implications of such intensification have taken various forms and include, for example, the elimination of residual vegetation cover, soil degradation, or water and land pollution (Briggs & Courtney 1989). The role of technology has been pivotal in this regard. For example, the introduction of the metal-tipped plough facilitated the use of heavier soils, whereas the introduction of chemical fertilizers and the genetic manipulation of crops have enhanced yields (Simmons 1993). Finally, large-scale mechanization of agriculture, as part of the development of a global agribusiness dominated by TNCs, has yielded further increases in production through the substitution of machines for human or animal labour (Harris 1978). The advance of technology has typically promoted more intensive permanent agriculture and it has also brought into cultivation areas previously unsuitable for permanent agriculture.

Because of the application of such intensive EM techniques, there is growing concern that modern agriculture is an unsustainable form of EM (Knickel 1990, Moffat 1992). Indeed, the worldwide quest for ever greater production is leading to diverse forms of environmental degradation. Soil erosion is seen by many as the most pressing concern arising from permanent agriculture (Blaikie 1985, Rozanov et al. 1990). Such erosion is complex, and can include soil compaction, nutrient loss, loss of organic matter, soil pollution, salinization, or changes in soil pH (e.g. acidification). In the UK, for example, erosion under cereal crops can be of the order of 30–95 tonnes per hectare in fields where hedges and other types of field boundaries have been removed (Pretty 1995). In aggregate, these processes can reduce agricultural productivity, and may even lead to irreversible environmental degradation, rendering soils useless for agriculture (Blaikie 1985, Stocking 1987).

Environmental degradation associated with unsustainable agrarian EM practices can also be less visible than soil erosion, but has no less severe implications. The impact of intensive permanent agriculture on hydrological systems is a case in point, and may include groundwater depletion, water pollution, and diversion of water (Briggs & Courtney 1989, Pitman 1992). Similarly, such agriculture may be linked to global atmospheric change (Parry 1990, Moffat 1992). The dramatic expansion of livestock numbers (there are 10 billion ruminants on Earth!) and the extension of rice cultivation have led to dramatic increases in greenhouse gases (e.g. methane), which in turn are associated with greenhouse warming (Clayton 1995).

These adverse environmental impacts raise the issue of whether intensive permanent agriculture as an EM practice can ever be sustainable. In earlier times, when such agriculture took place on a relatively limited scale, this issue was of purely local or regional significance. Since intensive permanent agriculture is now the pre-eminent agrarian EM practice in the world – in the process rapidly replacing rotational agricultural practices and hunting–gathering – this issue has assumed global significance. The environmental impact of permanent agriculture is the subject of considerable debate. On the one hand, Simon (1981) and North (1995) emphasize the productivity gains enjoyed under intensive permanent agriculture. Responding to neo-Malthusian writers, Simon points to the fact that farmers are able to produce enough food to feed the world's entire population. That hundreds of millions of people today are malnourished is seen as a reflection of unequal entitlement to food, and not as a result of the limitations of permanent agricultural practices themselves (Sen 1987). On the other hand, Schumacher (1973) and Pretty (1995) emphasize that intensive permanent agriculture can be antithetical to sustainable EM. Indeed, the increasing intensity and scale of environmental impacts associated with such agriculture have led some to speak of the "environmental contradictions of agriculture" (Buttel 1986, Bowler et al. 1992). These writers acknowledge the short-term gains associated with such EM practices, and emphasize the adverse environmental impacts that are likely to result in long-term environmental degradation and, ultimately, the collapse of this form of agriculture.

The discussion so far has concentrated on the central role of the farmer as an environmental manager. It is incumbent on the farmer to implement EM practices that help restore biomass and nutrients following agricultural production, so that environmental predictability is ensured. In this regard, intensive permanent agriculture as an EM practice is critically concerned with the nature and extent of both the taking from, and adding to, the environment. Modern agriculture may lead to environmental degradation because excessive biomass is extracted in a given area, forcing the farmer to add potentially environmentally harmful inputs such as chemical pesticides, herbicides and fertilizers in order to maintain production (the "technological treadmill"). As farming has become more capital-intensive, farmers have become integrated into the global political economy, such that the pressure to produce leaves little option but to maximize production (Goodman & Redclift 1991; see Ch. 6).

In the process of adopting such intensive EM practices, farmers now face an increased range of social and environmental uncertainties associated with the technological bases of modern agriculture. Just as industrial manufacturing has generated new uncertainties linked to the generation of pollution (see next section) in seeking to promote predictability *vis-à-vis* the physical environment so too has intensive permanent agriculture generated new uncertainties associated with the political economy of modern agriculture.

One response to the new EM uncertainties associated with intensive permanent agriculture has been the search for other EM practices that break out of the "productivist" ethos that has dominated world agriculture since the 1950s (Whitby & Lowe 1994). Increasing numbers of farmers implement organically based EM practices that attempt to replace extracted biomass by more environmentally friendly means. In this way, these environmental managers are striving to promote environmental predictability by reducing unwanted by-products (e.g. pesticides) added to the environment. Alternatively, farmers are seeking to use EM strategies that reduce production through less-intensive means of production or the withdrawal ("set-aside") of selected areas from production altogether (Buller 1992, Potter et al. 1993, Wilson 1994a, Hoggart et al. 1995). Although in the early stages of implementation, these strategies may hold the key to ensuring that maximum food and fibre production does not occur at the expense of long-term environmental sustainability. Farmer strategies are also influenced by the fact that modern intensive agriculture is based on fewer farmers controlling larger holdings. Although the amalgamation of farms is resulting in there being fewer farmers, the decisions of these more and more powerful environmental managers (i.e. owning larger farms) will play an increasingly important role in multi-layered EM (Morris & Potter 1995, Wilson 1997).

Intensive permanent agriculture is one of the main human interactions with the environment, but the role of farmers as environmental managers has been under-emphasized in many traditional accounts. In seeking to establish EM on a sustainable basis, a central task will be to consider the adverse environmental impacts associated with intensive permanent agriculture in a global capitalist economy. However, these impacts are dwarfed by the severe and extensive environmental degradation associated with industrial production.

Industrial production

Industrial production has resulted in intensifying environmental use. The implications of this process in terms of environmental degradation have today assumed global proportions. To the extent that environmental managers operate within a "risk society" (Beck 1992), a large part of such risk may be attributed to the effects of industrial production. More than ever before, uncertainty in EM is a by-product of industrial activity and the associated practices of environmental managers that often lead to environmental degradation.

Few would argue that industrial production enhances environmental predictability. Aside perhaps from scholars such as Simon (1981), the literature emphasizes the resource-depleting and polluting dimensions to industrialization. Whether it be taking from or adding to the environment, industrialization is often taken to be synonymous with environmental degradation (Devall & Sessions 1985, Lovelock 1995).

There are two central issues in considering the relationship between industrialization and EM. First, the ways in which industrialization has resulted in ever more intensive use of the environment, thereby contributing to increasing uncertainty in EM, need to be considered. Secondly, the implications in terms of which environmental managers are best positioned to undertake the necessary measures to reduce this uncertainty must also be addressed.

Industrialization as a type of human–environment interaction surpasses the environmental impacts of permanent agriculture in both scale and intensity. Indeed, the contemporary concern over the environment is largely based on the cumulative environmental effects of industrialization and associated changes in life-styles (e.g. consumerism). Industrialization is commonly seen as a radical break from preindustrial EM practices in the sense that it is based on extensive resource use and pervasive environmental pollution.

To understand the relationship between industrialization and increasing uncertainty in EM, the evolution of this increasingly important type of human–environment interaction needs to be outlined. Large-scale industrial development first occurred in late eighteenth-century Britain before spreading to America and other parts of Europe in the nineteenth century; in the twentieth century, large-scale industrialization is also occurring in such ELDCs as South Korea, India, Thailand or China (e.g. Bowonder 1986, Chia Lin Sien 1987, McDowell 1989). Typically, such development relates to the mass production of industrial and consumer products. Industrial production is based on the extensive use of energy supplies, originally in the form of coal, but since the early twentieth century increasingly through use of oil or even nuclear and hydroelectric power. It also requires a large consumer market for these goods. As many of these industrial products are not essential to human survival, yet require massive inputs of energy for their production, these goods are associated with consumption overpopulation (see p. 42), especially in EMDCs. Although industrial production is centred increasingly in ELDCs, consumption of the fruits of such production is as yet mainly concentrated in EMDCs. Further, industrialization has encouraged widespread urban development, as a large labour force was needed in the production process. The unprecedented concentration of people in large urban centres has, in turn, generated a whole new set of social and environmental uncertainties associated with urban environmental degradation that complicate the quest for predictability in EM (Berry 1990, Hardoy et al. 1992).

Industrialization has encouraged an unprecedented, if highly selective, increase in human material wellbeing. It is also the source of a multitude of environmental problems increasingly manifested at a global scale. These problems are a key factor behind the growing uncertainty facing all environmental managers, albeit unequally, in multi-layered EM. Industrial production is based on large-scale exploitation of resources, both potentially renewable (e.g. timber) and non-renewable (e.g. minerals and fossil fuels) (Rees 1990, Mather & Chapman 1995). As Redclift & Woodgate (1994: 56) argue, "industrial society has one of its most important bases in a fuel supply which, in terms of human time-spans, is in strictly limited supply". This taking from the environment is mirrored by adding to the environment through increased pollution, which is a direct by-product of the manufacturing process itself. Concurrently, urbanization linked to industrialization has resulted in concentrated human waste which further adds to environmental degradation (Brimblecombe 1987, Berry 1990).

Large-scale natural resource extraction has been a prerequisite for industrial production. This has been reflected in a quest for timber and minerals that has extended over at least the past two centuries. This quest has encompassed resources around the world, even in traditionally inaccessible areas such as Amazonia or the polar regions (Hecht & Cockburn 1989, Simmons 1993). This spatial expansion of resource exploitation has been accompanied by ever-more technologically sophisticated efforts to intensify exploitation in a given area (e.g. Judelson 1986). This is best illustrated with reference to the quest for oil, which increasingly encompasses efforts to tap offshore reserves. Intensive exploitation has also taken the form of extensive and highly destructive open cast mining (Miller 1994). In aggregate, these processes have often intensified environmental degradation through resource depletion, downstream pollution or toxic wastes (Simmons 1993). Such degradation has greatly increased environmental uncertainty in EM.

The ways in which industrial production pollutes the environment engenders most concern. Besides human health hazards, these emissions have a serious impact on the environment, with implications for the policies and practices of environmental managers (see Ch. 7). For example, acid deposition is contributing to *waldsterben* (tree death) in Germany, where it was first widely noted, but which has since been identified in many other forests of the world (Mohnen 1988, Mazurski 1990). Similarly, acid deposition in lakes has resulted in critical changes in water pH levels leading to the extermination of marine life, notably in northern Europe and North America (Mannion 1992). A potentially more serious environmental impact relates to global atmospheric warming. This phenomenon results from the increased concentration of gases that trap heat in the atmosphere (Clayton 1995). Uncertainty surrounding the precise impact of global warming is chronic, rendering predictions of future environmental change highly problematic. Similar doubts surround the precise environmental impact of industrial effluents on oceans. Yet, research shows that pollution levels (e.g. sewage, industrial effluents, oil) have increased dramatically; for example, lead levels have quadrupled in some parts of the oceans since 1700 (Jickells et al. 1990).

Whether it be in terms of resource depletion, pollution, or even altered consumer life-styles, industrialization is linked to many of the environmental problems with which environmental managers must deal today. Uncertainty in EM has, thus, assumed global proportions. This, in turn, renders the task of enhancing predictability in EM increasingly difficult.

At the same time, industrialization has created a need for EM on a scale and complexity that only the state and other powerful international managers (e.g. TNCs) are in a position to undertake. Industrial environmental impacts reinforce the importance of the state in selected aspects of EM. The ensuing uncertainty requires a level of institutional and financial capability that is beyond many non-state environmental managers. The nature of such uncertainty reflects both the central role of the state in promoting industrialization in the first place (see below), and its critical function in any initiative that aims to alleviate the environmental problems that ensue from this production. Just as some farmers are seeking to substitute environmentally benign inputs to the production of food for existing damaging inputs, so too state environmental managers may seek to mitigate environmental damage resulting from industrial production through implementation and enforcement of appropriate policies. However, powerful non-state environmental managers, such as TNCs, international financial institutions or environmental NGOs, are increasingly able to address environmental problems related to industrial production (Pearson 1987, Hurrell 1994, Wapner 1995). Conversely, many small or poverty-stricken states lack the means to implement sustainable EM policies that combat environmental degradation associated with industrialization.

This discussion of industrial production highlights how intensifying human use of the environment has increased the complexity surrounding all dimensions of uncertainty in EM. It has also highlighted how such production (and associated environmental degradation) may strengthen the position of certain environmental managers in multi-layered EM. Yet, what still needs to be considered is why the human impact on the environment has so dramatically intensified in recent centuries, thereby providing the frame of reference within which contemporary EM must operate. The following discussion will highlight the growing historical significance of the modern state as a key type of environmental manager – a process intimately associated with the rise of capitalism as the predominant form of economic activity in the world.

The political economy of environmental management

Traditional EM literature has often neglected the historical context within which contemporary EM operates (for a notable exception, see Pepper 1984). Attention focuses on the details of environmental problems and state EM techniques at the expense of a wider appreciation of the political economy of EM. Yet the latter is central to understanding why uncertainty has increased for state and non-state environmental managers acting in multi-layered EM. To appreciate the political economy of EM is, in turn, to consider the EM impact of a globalizing capitalist system whose development has been helped by the emergence of the state as a key environmental manager in human–environment interaction.

The role of capitalism in such interaction is especially critical. That role is considered in greater detail in Chapter 6 in relation to a discussion of the market and EM. The objective here is to situate an understanding of the development of multi-layered EM in the context of the historical evolution of capitalism as a system of production.

The link between capitalism and social and environmental wellbeing is controversial. Indeed, some have even asked whether capitalist development and the survival of humankind are mutually contradictory processes (Sklair 1994). Capitalism developed in Europe after the Middle Ages as a distinctive and increasingly powerful form of economic organization. In contrast to previous forms of economic activity based on reciprocity and redistribution, capitalism developed as a dynamic process in which the quest for profit was a paramount concern (Polanyi 1957, Peet 1991). As Chapter 6 explains, capitalism is based on intensifying environmental use, a point illustrated here in an historical context.

A continual search for cheap resources has been a hallmark in the development of capitalism. The role of capitalists was crucial in European expansion overseas from the fifteenth century (Wolf 1982). However, it was the growing industrialization of Europe and North America in the nineteenth century that intensified the overseas quest for raw materials (as well as markets for manufactured goods). Most evidently, colonialism led to non-sustainable timber extraction resulting in widespread deforestation in colonial territories. The impetus behind such exploitation by European colonial powers, in turn, reflected timber shortages in Europe that had resulted from non-sustainable EM practices. This process was replicated in many ELDCs. For example, following large-scale forest clearance prior to the nineteenth century in Great Britain (Hoskins 1955, James 1981), the British pursued a strategy of timber exploitation in their colonies, notably in India, Burma and Australasia (Guha 1989, Roche 1990, Memon & Wilson 1993). Such extraction was conducted initially according to laissez-faire principles, with capitalists given a free hand to extract as much timber as they liked without regard to the future of the forests (Tucker & Richards 1983, Williams 1989, Bryant 1994b). The result was widespread deforestation. For example, in northern India in the 1860s, and in New Zealand in the 1890s, thousands of square kilometres of forest were destroyed in a matter of only a few years.

At a more general level, the impact of capitalism was associated with radically altered EM practices around the world. Just as during an earlier era in Europe, small-scale environmental managers in colonial territories found that EM practices hitherto utilized were no longer appropriate in a context of integration in the global capitalist economy (*The Ecologist* 1993). At its most extreme, this point is illustrated by farmers who left the land permanently and migrated to cities. These people lost their status as environmental managers because, as a result of this move, they no longer actively and self-consciously manipulated the environment for their livelihood.

The introduction of cash crop production in rural areas, meanwhile, led to a fundamental restructuring of the EM policies and practices of many local-level environmental managers. In most precolonial societies, EM practices were predominantly (but not exclusively) geared towards subsistence needs or local markets. Integration in the global capitalist economy changed all that. Thereafter, EM was largely based on exploitation of the environment for profit, either by smallholders or by large-scale capitalist firms such as the Bombay–Burmah Trading Corporation or the United Fruit Company, precursors of today's TNCs (Wolf 1982, Rush 1991, Bryant 1997). In this manner, colonialism not only resulted in massive environmental change; perhaps more significantly, it also disrupted existing practices in multi-layered EM. For example, colonial commercial logging practices invariably undermined pre-existing EM practices of shifting cultivators; the former were not necessarily sustainable. On the contrary, the new practices linked to capitalism were often highly damaging to the environment. Thus, the advent of capitalism under colonialism resulted in the substitution of a set of practices that were more damaging to the environment than was hitherto typically the case, thereby increasing uncertainty in EM. Yet, such uncertainty was not equally distributed. For example, poor farmers and shifting cultivators were more adversely affected by the consequences of environmental degradation than were their wealthier counterparts.

The emergence of capitalism was, therefore, often linked to widespread environmental degradation, but occasionally it also led to EM practices based on the promotion of sustainable EM policies. At the same time as the colonial powers encouraged widespread forest clearance, for example, they also introduced scientific forestry in selected commercially valuable forests. Scientific forestry, a technique first developed in Germany at the end of the eighteenth century, is a system designed to ensure commercial timber extraction according to the principles of sustained yield (Heske 1938, Wiersum 1995; see Ch. 2). Often, these attempts were shortlived and covered only relatively small areas, compared to the large areas of forest destroyed for the production of food and fibre in a globalizing capitalist economy (LeHeron 1988, Rush 1991). For example, in British Burma, the Forest Department created a network of reserve forests which at their greatest extent covered about 15 per cent of the national territory. However, such forest protection in aid of long-term commercial teak extraction occurred in a context in which all low-lying forests suitable for permanent agriculture were cleared in the late nineteenth and early twentieth century (Adas 1974, Bryant 1997).

Such massive environmental changes – manifestations of a territory's integration in the globalizing capitalist economy – have usually continued since the ending of formal colonial rule. Yet, with very few exceptions, independence did not result in the termination of economic links to the global capitalist economy. On the contrary, patterns of economic activity first elaborated under colonial rule often were intensified in the postcolonial era. For example, many ELDCs still rely on natural resource exploitation as the basis of national economic activity, just as they did in former colonial times (Watts 1983a). The chronic indebtedness of many ELDCs to EMDC states, banks and international financial institutions since the early 1980s has forced most ELDC states further into the capitalist market. This process has intensified unsustainable EM practices.

At a general level, the development of capitalism has encouraged unsustainable EM practices in multi-layered EM. Non-state environmental managers such as farmers responded to the new incentives and pressures of capitalism by adopting EM practices based on ever-more intensive environmental exploitation. To take but one example: in New Zealand, pioneer farmers in the late nineteenth century cleared large tracts of old-growth forest for sheep and cattle farming linked to meat exports to the UK. Such practices resulted in large-scale soil erosion on slopes now denuded of protective forest cover (Cumberland 1941, Wilson 1993).

A recurring pattern emerged as non-state environmental managers cleared forest and intensified production as part of the switch from subsistence, or local small-scale market production, to larger-scale profit-orientated production for the capitalist market. These environmental managers often received the strong backing of the state. Indeed, capitalist production could scarcely have been introduced on a global scale without the active intervention of states. In the process, the state as an environmental manager assumed a central role in multi-layered EM. Capitalism, therefore, has been based on the integration of far-flung territories, resources and peoples, but appears to have needed the state to facilitate its expansion (Hall 1986, Mann 1986).

The specific characteristics of the state as an environmental manager in multi-layered EM are examined in detail in Chapter 5. Here, the role of the state in relation to capitalism needs to be briefly described, and six issues relating to the emergence of the state as an environmental manager are considered.

First, the state was crucial to the development of capitalism in that it was the institution that became responsible for the provision of public or collective goods. Such goods came to include a common currency, national defence, education, and the maintenance of law and order, especially the protection of private property (Johnston 1989). These goods are important in terms of EM in that the state's provision of such goods created an appropriate climate for the accumulation of capital, a large part of which related to the exploitation of the environment by capitalists.

Secondly, the state played a direct role in facilitating the extraction of natural resources essential to economic activity under capitalism. That role included, for example, the provision of tax and other financial incentives to businesses to conduct economic activities such as logging, mining, cattle ranching and agriculture (Rush 1991). Capitalism did not integrate territories around the world naturally. Rather, such integration required an active state role in encouraging capitalist enterprise into hitherto "marginal" regions. Besides financial incentives, states also provided political "security" for potential investors as the state used its coercive powers to suppress opposition to the advent of capitalism in a given territory. States also typically sponsored massive infrastructural schemes to facilitate the extraction of natural resources. Such schemes included, for example, road and rail networks, but also the construction of dams, irrigation channels and power stations designed to provide water and electricity. In Burma, as elsewhere in the colonial world, the state implemented policies designed "to facilitate the movement of labour and export products and to make cultivation of empty lands possible" (Adas 1974: 35).

Thirdly, the state also attempted selectively to conserve environmental resources important to the globalizing capitalist economy. For example, many colonial powers sought to identify and protect key commercial timbers such as teak, cedar and kauri in diverse parts of Asia or Australasia. They did so through the creation of specially protected state "reserved" forests in which intensive EM based on the idea of sustained yield was often attempted (Guha 1989, Peluso 1992, Memon & Wilson 1993, Bryant 1997). In some cases, that attempt brought the state into direct conflict with some capitalists anxious to maximize short-term extraction and profits. The ultimate goal of such state EM practices was long-term resource use linked to the capitalist economy, but this often served to mitigate potential conflict between state agencies and capitalists.

Fourthly, and following from the previous point, the role of the state was not solely to facilitate capitalism. As Skocpol (1985) and Mann (1986) note theoretically, the state is much more than simply an agent of capitalism, but it has its own sources of power that derive from its unique position at the intersection of the national political order and the interstate system. This point has implications for the state's role as an environmental manager in multi-layered EM. As noted, state policies often facilitate capitalist activities, but from time to time they have also sought to further other objectives. For example, beginning in the nineteenth century, the creation of a network of national parks with the primary aim of protecting distinctive remnant habitats and wildlife occurred in many cases against the backdrop of business opposition (Hall 1988, Williams 1989, Shultis 1995).

Fifthly, and reflecting the state's diverse political and economic responsibilities, the role of the state in multi-layered EM has increased over the centuries. Under capitalism the state has assumed a particularly prominent position in regional, national and international EM decision-making (Johnston 1989, Hurrell & Kingsbury 1992). As part of that process, large bureaucracies developed to manage diverse environmental resources controlled directly by the state, and to regulate the EM practices of non-state environmental managers, issues examined in more detail in Chapters 5 and 7. What is important to note here is simply the sheer growth in the financial and human resources dedicated by the state to EM matters over time, and the associated increase in the power of most states to influence practices in multi-layered EM. As Chapter 5 notes, although the growing power of such non-state environmental managers as TNCs, international financial institutions and environmental NGOs today may be associated with a declining capacity of the state to control EM policies and practices.

Finally, the promotion of capitalism, as well as the state's own interests, has been associated with policies and practices that have had an adverse social and environmental impact on many small-scale environmental managers. As part of the expansion of natural resource exploitation and infrastructure development projects, states have displaced millions of hunter–gatherers, shifting cultivators and peasants from their lands, thereby disrupting existing local EM practices. Such disruption has tended to increase the uncertainty faced by these local environmental managers. In this manner, state efforts to integrate hitherto autonomous environmental managers politically and economically have often been associated with ever more complex strategies conditioning the EM policies and practices of non-state environmental managers. State policies may take the form, for example, of proscriptions on forest clearance, specified limits to pesticide use, or maximum allowable kills for endangered species, and have frequently led to conflict between state and non-state environmental managers. Yet, the growth of such conflict in EM is related in complex ways to the growth of state powers and an associated push to expand capitalist economic production around the world.

Conclusion

This chapter has shown that the human impact on the environment has increased over time because of population growth and intensifying human use of the environment. These changes in human–environment interaction have, in turn, increased the uncertainty facing most environmental managers at the local, national and global scales. The need to appreciate the role of the globalizing capitalist economy and the empowerment of states over time in this process was also emphasized.

More people has meant more environmental managers operating in multi-layered EM. In some cases this has resulted in an increased competition for scarce environmental resources. Further, the factors that all environmental managers need to take into account in their decision-making have increased as EM itself has become more complex.

On the one hand, intensifying human use of the environment has been associated with the growth of a wide range of environmental problems that often have constrained the ability of environmental managers to promote predictability in EM. For example, whereas forestry officials in the eighteenth century may have been concerned with the commercial and environmental implications of unsustainable forestry, their contemporary counterparts have also to worry about the impact of global environmental pollution (e.g. acid rain) on forest survival.

On the other hand, EM as a multi-layered process has become more complex over time, further complicating the efforts of environmental managers to pursue predictability through their EM policies and practices. The discussion of the different types of human–environment interaction noted the environmental implications of the shift from relatively low-intensity EM practices to high-intensity EM practices over time. Part and parcel of this historical shift in human–environment interaction has been the growing importance of new environmental managers. The role of the state was emphasized in this regard, notably in relation to the spread of capitalism. Yet, recent decades have also witnessed the emergence of TNCs, international financial institutions and environmental NGOs as new types of environmental managers (a subject considered in subsequent chapters). The overall effect has been increasing complexity – new types of environmental managers interacting with long-established environmental managers – and in the process adding new layers to multi-layered EM.

The historical record is, thus, one of increasing uncertainty for most environmental managers in a context of intensifying human use of the environment. But uncertainty in EM does not end there. It is further complicated by the often ambiguous nature of environmental attitudes, worldviews, and discourses that underpin EM practices in multi-layered EM, the subject of Chapter 4.

CHAPTER 4
Environmental worldviews, attitudes, and discourses

To understand the policies and practices of environmental managers operating in multi-layered EM is to appreciate partly how environmental worldviews, attitudes and discourses condition EM decision-making processes. As Chapter 3 noted, those processes may be influenced by political and economic forces, but they are also culturally constructed, adding potentially greater uncertainty to an already complex EM situation. EM practices can certainly not simply be "read off" from cultural attributes such as environmental worldviews, attitudes and discourses. Yet, to understand why environmental managers act as they do, it is critical to understand how they perceive the environment, their relationship to it, and their interactions with each other.

Let us start with an assessment of the "ecocentric" and "technocentric" worldviews held by different environmental managers. Arguably, a more precise measure of the practical significance of cultural conceptions of human–environment interaction is that of the environmental attitudes of different types of environmental managers. However, those attitudes are not isolated from one another; rather, they are expressed through environmental discourses as environmental managers seek to assert and justify "appropriate" forms of human–environment interaction.

Environmental worldviews

How the management of the environment is perceived has been linked to culturally constructed environmental worldviews. Many scholars have attempted to explain the complexities of different environmental worldviews (e.g. O'Riordan 1981, Cotgrove 1982, Pepper 1984, Eckersley 1992, O'Riordan 1995b). The most useful approach in terms of an appreciation of the impact of worldviews on EM practices is that of O'Riordan (1981, 1995b), who distinguishes between "ecocentric" and "technocentric" worldviews. This framework provides a useful means to situate at a general level different types of environmental managers according to their beliefs about human–environment interaction.

Ecocentrism is a worldview in which predictability is sought by giving priority to environmental conservation over exploitation. Technocentrism, in contrast, emphasizes the pursuit of predictability through human ingenuity (i.e. technological innovation) and intensive use of the environment. These worldviews are conditioned not only by perceptions of the environment but also "are embedded in a host of other social, political, and economic outlooks" (O'Riordan 1995b: 5).

Ecocentrics typically advocate that humans must live within well defined environmental bounds. Writers such as Devall & Sessions (1985), Naess (1989) and Sessions (1994) emphasize the need to manage the environment so that resources are not depleted. They criticize contemporary EM practices based on economic growth and global capitalism, which are seen as the main contributors to environmental degradation. However, there are diverse perspectives within ecocentrism (Pepper 1993, Jacob 1994). For example, radical ecocentrism – as reflected in work in "deep ecology" (e.g. Devall 1988) or the "Earth First!" movement (Foreman & Haywood 1988) – calls for nothing less than a revolutionary transformation of human–environment relations and that absolute priority be given to environmental conservation. As Karshenas (1994: 733) observes, deep ecologists "expect the whole world to return to preindustrial, rural life-styles and standards of living". In contrast, less radical ecocentrics, as typified, for example, by *The Ecologist* (1972), Schumacher (1973), or Friends of the Earth (1994a), assert the importance of local-level EM practices based on environmental conservation, but attained through the reform of the existing policies and practices of diverse environmental managers.

Technocentrics would dismiss such arguments as hopelessly idealistic and naïve. As reflected in the works of Simon (1981), the World Bank (1992), and Karshenas (1994), technocentrics do not dwell on environmental problems, but emphasize technological and market-based EM solutions. In contrast to the ingrained pessimism of ecocentrics, technocentrics are confident that traditional state environmental managers will resolve even the worst environmental crises. This will be achieved through the entrenchment of human wellbeing rather than "pristine" environmental quality (Lorrain-Smith 1982). There is again a great diversity of perspectives. Extreme technocentrics or "cornucopians", such as Simon (1981), Simon & Kahn (1984) and Beckerman (1995), suggest that there are no environmental crises at all. In contrast, moderate technocentric views, notably espoused by the World Bank (1992), Barbier (1987) and Pearce et al. (1989), acknowledge the existence of environmental problems but are confident that technological solutions can be developed in order to reform existing EM policies and practices. Indeed, as Garlauskas (1975: 194) asserts, "environmental management can be applied as a systematic tool to preserve this equilibrium between man [sic] and the environment" (see also Davis & Nanninga 1985). Indeed, and as Miller (1985, 1993) argues, some authors equate EM itself with moderate technocentrism.

Differences between the ecocentric and technocentric worldviews are perhaps most evident over the question as to whether there is a contemporary environmental crisis (Myers & Simon 1994). According to many ecocentrics, Earth has finite resources, and increasing production, consumption and human population will inevitably destroy the environment. In this context, Lovelock (1995) has developed the Gaia hypothesis, which suggests that the Earth is a single living entity in which the survival of any one species – including humans – is not guaranteed. In this view, human-induced environmental degradation may lead ultimately to the elimination of human life on Earth. Unless environmental managers introduce appropriate EM practices, environmental crises spell the end of humankind. To avert such crises, ecocentrics emphasize that humans must adapt to extreme environmental uncertainty through radically altered life-styles and local-level EM practices (Schumacher 1973). A central implication of the ecocentric worldview is that centralized EM practices and the global capitalist economy must be abandoned if sustainable EM is to be achieved.

In contrast, technocentrics assert that human ingenuity through innovative EM practices will avert any possible environmental crisis. Indeed, Simon (1981), Simon & Kahn (1984), and Beckerman (1995) argue that the adoption of innovative policies and practices by diverse environmental managers will enable humanity to continue to enjoy long-term economic growth, without impairing the Earth's life-support systems. For this reason, Simon & Kahn (1984) argue that the very notion of a crisis – and all that follows from it – is "dead wrong" (cf. Adams 1990). They point to the fact that past resource shortages, such as fuelwood crises in seventeenth-century England, have not only been overcome by the use of coal, for example, but have in the end left humanity better off than before, thereby, they argue, confirming that innovative EM practices can overcome environmental uncertainty. Thus, the technocentric worldview emphasizes "problem solving" as the central way in which to manage the environment, but requires that detailed environmental data be available. Barbier (1987) and Pearce et al. (1989), for example, suggest that through the provision of adequate knowledge – and notably the costing of environmental goods – to state environmental managers, it will be possible to modify existing EM practices so that environmental degradation can be stopped.

These environmental worldviews form part of a much broader history of cultural and economic development (Glacken 1967, Tuan 1972, Simmons 1993). For example, the spread of the Judeo-Christian religion led to the view that humans are separate from nature, and that nature exists solely for human use. This view highlights a technocentric approach that may have had a powerful influence over EM over the centuries. For example, White (1967: 1204) suggested that Christianity encouraged the belief that humanity is "superior to nature, contemptuous of it, willing to use it for [human's] slightest whim". Passmore (1980) asserts that this ethos of domination over nature even predates Christianity and is rooted in ancient Greek and Roman civilization. It can be argued that, in Western societies, Graeco-Roman and Christian notions of human mastery over nature may be an important strand in technocentric approaches to EM.

As Passmore (1980) points out, Christianity has also been used to sustain the ecocentric worldview. Indeed, the notion of human "stewardship" over nature is a recurrent theme in the New Testament. As Black (1970: 48–9) argues in this regard, if nature was made in God's image, then the implication for EM is that humankind should act "in a responsible way in relation to the lower order of creation in the same way as God acts upon man [sic]". This stewardship ethos is a central part of an ecocentric worldview, but may also imbue the EM practices of many environmental managers (cf. Simmons 1989, 1993).

Outside the Western cultural context, it has been argued that other cultures have provided propitious conditions for the development of an ecocentric worldview. For example, the role of Buddhism in many Asian societies has been associated with an inclination towards a symbiosis of humans with nature (Callicot & Ames 1989). However, as recent research has shown, Asian cultural contexts have not necessarily resulted in sustainable EM practices (Bruun & Kalland 1994).

Another example highlighting the contextual influences on the development of environmental worldviews relates to the ascendancy of capitalism around the world. Chapter 3 discussed the material impact of capitalism on the environment, which also partly conditioned how environmental managers perceive the environment. Capitalism has encouraged the ascendancy of a technocentric worldview in the policies and practices of many environmental managers that has often been associated with widespread environmental degradation (see Ch. 6). Yet, environmental degradation and technocentrism are not necessarily synonymous with capitalism (Walker 1989), although capitalism undoubtedly has been a particularly potent force in this regard. Capitalism cannot be static. Driven by ever-changing technologies, capitalism "cannot sustain an equilibrium with its environment" (Johnston 1989: 49). It is for this reason that ecocentrics criticize both capitalism and technocentrism, as the latter often leads to EM practices that cause environmental degradation and, therefore, increased uncertainty in EM. Proponents of the technocentric worldview respond that it is precisely the accumulation of capital that gives environmental managers the ability to reduce such uncertainty in EM (e.g. Barbier 1987).

This discussion of ecocentric and technocentric worldviews has shown two contrasting ways of perceiving and understanding human–environment interaction. The importance of this discussion is threefold. First, it shows how the policies and practices of environmental managers may be influenced by broader cultural constructs. All environmental managers operate to a greater or lesser extent in keeping with an environmental worldview. That worldview provides an essential frame of reference for environmental managers as they devise and seek to implement environmental policies. At a general level, the key types of environmental managers referred to in this book may be associated with ecocentric or technocentric worldviews. States, TNCs, international financial institutions and large-scale farmers may broadly be linked to the technocentric worldview that now prevails in global political and market relations. In contrast, hunter–gatherers, shifting cultivators, environmental NGOs and many small-scale farmers may broadly be associated with the ecocentric worldview.

Secondly, an environmental worldview provides only a broadbrush understanding of an environmental manager's perception of human–environment interaction. On the one hand, different individual environmental managers of the same broad type around the world may adhere to a technocentric or ecocentric worldview, depending on specific cultural contexts (e.g. religion) or economic circumstances. On the other hand, the policies and practices of an individual environmental manager may themselves reflect the combined influence of ecocentric and technocentric worldviews. In effect, this situation is indicative of the tensions and ambiguities surrounding appropriate patterns of human–environment interaction that may be represented in individual EM practices.

Thirdly, and despite the complex ways in which individual environmental managers may absorb ideas from ecocentric and technocentric worldviews into their own outlooks, it may be possible to suggest a broad historical trend in the development of societal environmental worldviews. The history of intensifying human use of the environment (see Ch. 3) may be largely associated with the triumph of the technocentric worldview in society closely associated with the spread of global capitalism. In contrast, disillusionment with the adverse social and environmental effects of capitalism may be promoting a shift in contemporary societal outlooks towards a more ecocentric worldview (Inglehart 1977, O'Riordan 1995b).

This discussion of environmental worldviews has shown the potential influence of broad cultural forces and ideas on the policies and practices of environmental managers. To appreciate the specific ways in which environmental worldviews may impinge on the day-to-day activities of environmental managers, it is important to explore further the cultural dimensions to multilayered EM through an analysis of the environmental attitudes of environmental managers.

Environmental attitudes

We now examine the environmental attitudes of different types of environmental managers, and the practical significance of these attitudes in terms of EM policies and practices. The formation of environmental attitudes, and the possible links between attitudes and behaviour, are still subject to debate (Ajzen & Fishbein 1980, Gray 1985, Ajzen 1988, 1991, Eagly & Chaiken 1992,

Steel 1996). It is important here to highlight generally the interrelationship of environmental attitudes, intensifying human use of the environment, and increased social and environmental uncertainty in EM.

Although environmental attitudes are often complex, they can nevertheless be generally classified along a spectrum ranging from highly utilitarian to staunchly conservationist (Crandall 1980, Kellert 1984). These attitudes can be related to the environmental worldviews discussed above: utilitarian attitudes with the technocentric worldview, and conservationist attitudes with the ecocentric worldview. At their most extreme, the difference between utilitarian and conservationist attitudes is striking. Environmental managers with highly utilitarian attitudes see the environment predominantly as a source for economic wellbeing and profit (a "utility"). The environmental attitudes of some farmers with large holdings in eighteenth-century England are a good expression of such attitudes. In order to raise large sums of money, many such environmental managers practised unsustainable EM. Indeed, as described by the second Earl of Carnarvon, trees were regarded as "an excrescence of the Earth, provided by God for the payment of debts" (Thomas 1983: 200). Utilitarian attitudes, therefore, may be associated with unsustainable EM practices that contribute to environmental uncertainty.

In contrast, environmental managers who adhere to strongly conservationist attitudes profess a deep reverence for the environment. They consider unspoilt nature an essential source of human survival. In many cases, such attitudes represent a reaction to the utilitarian attitudes just described. In eighteenth century England, while trees were felled indiscriminately by utilitarian landowners, others decried such wanton destruction of the country's natural beauty. As William Gilpin remarked, trees could not be reduced to their commercial value alone, but were "the grandest and the most beautiful of all the productions of the Earth" (Thomas 1983: 213).

Environmental attitudes may have a direct bearing on the policies and practices of environmental managers. This issue is explored in greater detail below, in relation to an analysis of the environmental attitudes of different types of environmental managers. First, however, it is useful to compare at an aggregate level how environmental attitudes have differed between preindustrial and industrial societies. Such a discussion provides further clarification of the cultural sources of intensifying human use of the environment.

Preindustrial and industrial environmental attitudes

It is commonly suggested that the environmental attitudes of preindustrial societies (e.g. hunter–gatherers) have been based on a "deep respect" for the environment (Denslow & Padoch 1988, Guha 1989, Durning 1993). Yet, if environmental attitudes prevalent in these societies may have served to prevent environmental degradation, sustainable EM was not always the result.

Preindustrial societies have generally attached more spiritual significance to their environments than industrial societies (Hooper 1981, Young 1992,

Beinart & Coates 1995). As expressed through rituals, oral histories, or cultural artefacts (e.g. masks, totem poles), preindustrial environmental managers have articulated their often complex relationships with wildlife, plants, landscape features and other parts of the environment (Blackburn & Anderson 1993, Reichel-Dolmatoff 1996). The fact that omnipotent spirits were said to inhabit certain parts of the environment – and that these spirits would punish all those who degraded the environment – is often taken as evidence that preindustrial environmental managers tended to adopt conservationist environmental attitudes.

In many cases these attitudes were based less on a respect for the natural world than on a fear of the potential perils of that world (e.g. dangers associated with hunting). As such fears declined through familiarity, the prohibition of over-exploitation of the local environment may too have declined (Boomgaard 1994). Further, colonial rule resulted in the introduction of new practices that violated previous spiritual sanctions on environmental mismanagement but without spiritual "punishment". For example, in mid-nineteenth century colonial Burma, shifting cultivation, as practised by the Karen indigenous people, was embedded in a complex web of permitted and prohibited environmental practices associated with conservation-orientated spirit (*nat*) worship. This EM practice was substantially altered with the advent of new environmental policies based on scientific forestry, which prompted a more utilitarian relationship between shifting cultivators and the forest (Bryant 1994a).

In many cases, preindustrial societies showed elements of both conservationist and utilitarian attitudes to the environment. Although some preindustrial environmental managers may have protected their environment, such respect may have been tempered by the necessity of sustaining a livelihood. For example, between the tenth and nineteenth centuries, the Maori of New Zealand developed a rich culture in which many individual groups and tribes developed a range of taboos and norms that effectively conditioned their EM practices (Fleet 1984). Some parts of their environment were imbued with *mana* (spiritual energy flows), such that these areas were subject to specific taboos often preventing over-exploitation (Hooper 1981). The forests, in particular, were associated with strong *mana* and taboos that resulted in the strict conservation of selected, and especially spiritually significant, areas. For example, Waipoua Forest in the extreme north of New Zealand was long protected by local Maori groups because some of the large kauri trees in that forest were sacred and could not be felled (e.g. *Tane Mahuta* the "God of the Forest"; McGregor 1948, Fleet 1984). For these groups, over-use or depletion of resources may have posed a severe threat to their survival in that there were fewer alternative strategies available to them than to industrial societies.

Whereas some Maori groups conserved certain spiritually imbued parts of the environment, other groups also ruthlessly exploited the environment as part of their EM practices (see Ch. 3). Not only did tribes during the early "moa-hunter" phase of Maori culture (*c.* AD 750–1300) contribute to the extinction of

40 moa bird species, they also cleared large areas of biologically diverse forests for hunting (Cumberland 1961). Indeed, it is estimated that they destroyed at least 30000 km^2 of forest, representing almost one-sixth of the original forest cover. In this way, the moa-hunter tribes destroyed their livelihood basis by increasing environmental uncertainty, resulting in their eventual demise (Cumberland 1961, McGlone 1983). Such "utilitarian" EM practices clearly show the complexity of the environmental attitudes of preindustrial environmental managers.

Notions of the "noble" savage living in total harmony with their environment, therefore, are a myth (Putz & Holbrook 1988). Although environmental attitudes of some preindustrial environmental managers led to a recognition of the need to conserve the environment, others adopted a more utilitarian approach grounded in the immediate satisfaction of needs and, thus, possibly undercutting their livelihood bases in the long term. This has recently been reiterated by Harvey (1993: 29), who argues that

> indigenous groups can . . . be totally unsentimental in their ecological practices. It is largely a Western construction, heavily influenced by the romantic reaction to modern industrialism, that leads many to the view that they were and continue to be somehow "closer to nature" than we are . . . even when armed with all kinds of cultural traditions and symbolic gestures that indicate deep respect for the spirituality in nature, they can engage in extensive ecosystemic transformations that undermine their ability to continue with a given mode of production.

In recent years scholars have nonetheless emphasized the potential benefits of reviving preindustrial environmental attitudes as the basis of sustainable EM (Gadgil et al. 1993). As a result of the growing environmental problems associated with industrial development and the spread of capitalism around the world (see Ch. 3), it is argued that contemporary societies could have much to learn from preindustrial societies about the appropriate attitudes to take towards the environment. The assumption here is that preindustrial societies are closer to the natural environment and are more sensitive to environmental perturbations than are industrial societies. Yet, "indigenous or pre-capitalist practices are not . . . necessarily superior or inferior to our own just because such groups espouse respect for nature rather than the modern . . . attitudes of domination or mastery" (Harvey 1993: 30).

One of the most significant historical changes in the environmental attitudes of many environmental managers related to the spread of industrialization around the world. The need for natural resources in manufacturing, and the pollutants that are the unwanted by-products of that process, have been a vivid manifestation of a utilitarian-based quest to industrialize. Many have argued that industrial development represented a dramatic departure from preindustrial practices and attitudes towards the environment (Simmons 1989, 1993, Miller 1994).

Industrial development resulted in major changes from preindustrial societies in how the environment was used and managed. It has also been associated with new ways of interpreting, relating to, and appreciating the environment by environmental managers, which were based increasingly on utilitarian attitudes. The exact relationship between the advent of industrial society and the growing prominence of utilitarian attitudes is far from straightforward and needs to be understood in relation to sweeping changes that affected all types of environmental managers operating within an increasingly multi-layered EM.

An initial important change related to the separation of a large proportion of the population from direct daily interaction with the environment. The mass movement of people from rural areas to rapidly growing urban areas was associated with the transformation of many erstwhile environmental managers into purely environmental users – that is, migration in some cases involved a loss of status as environmental managers (see Ch. 1). Many farmers, for example, whose livelihood had previously derived primarily from "active and self-conscious manipulation" of the land (farming) became urban workers and consumers with often only the remotest of connections to the environment (Bowlby & Lowe 1992). This process was associated with a gradual transformation of environmental attitudes among urban environmental users, which ultimately had important implications for EM as a multi-layered process (see below).

Industrialization and the associated mass movement of people from country to city also had important implications for environmental managers who remained in their traditional occupations. Industrialization was associated with the transformation of agrarian practices that was often also linked to altered environmental attitudes. Such factors as the consolidation of landholdings, mechanization, increasing reliance on chemical inputs (e.g. pesticides), and greatly increased production for the capitalist market – all had a combined effect of encouraging a more "businesslike" utilitarian attitude on the part of farmers towards the environment (Hawkes 1978, Knickel 1990, Ward & Lowe 1994). The point here is not to suggest that preindustrial agriculture was necessarily based on conservationist attitudes, or that industrial agriculture has been completely incompatible with elements of a conservation-orientated attitude, but it is indisputable that the advent of industrial agriculture around the world has been linked to a more thoroughgoing utilitarian attitude (see below).

Industrialization was associated with further changes that altered the very structure of multi-layered EM, with associated implications for environmental attitudes. To begin with, industrial development was linked to the growing importance of nation states, which gave the state a much larger role in EM than was hitherto the case. As Chapter 3 noted, a principal objective of the state has been to support the spread of capitalism. State environmental managers have been largely characterized by utilitarian environmental attitudes based on the maximization of human use of environmental resources (Johnston 1989). These environmental managers have also sought to inculcate utilitarian attitudes among some environmental managers who directly interact with the environment (e.g. farmers, shifting cultivators), thereby promoting a broader societal shift in attitudes.

Industrial development was also associated with the rise of new types of environmental managers with distinctive environmental attitudes. On the one hand, it gave rise to the development of large business firms, and ultimately TNCs, which have had an increasing impact on EM worldwide. These firms have been characterized by strongly utilitarian attitudes linked to a quest for profit (Eden 1994). On the other hand, industrial development has been associated with the growth of a large and somewhat diffuse "conservation movement" which has manifested itself increasingly in the form of environmental NGOs with staunchly conservationist attitudes (McCormick 1995). As discussed in more detail below, the development of the conservation movement was associated with a largely urban and middle-class population that "revolted against the "excrescences" of industrial capitalism . . . [as] symbolized in the 19th-century city" (Pepper 1984: 84). This conservation movement, and associated conservationist attitudes, did not begin to have a major impact on environmental attitudes until well into the twentieth century, at a time also when the role of the NGO as an environmental manager first became prominent. As a result, industrialization was initially at least largely associated with the triumph of utilitarian attitudes in society. This process stood in sharp contrast to the predominantly (although not exclusively) conservationist attitudes of preindustrial societies.

The preceding discussion focused on the environmental attitudes of two different forms of social organization and production in order to relate the discussion in Chapter 3 of intensifying human use of the environment to changing environmental attitudes. We now need to explore the attitudinal basis of EM practices further, which is best achieved through an analysis of the attitudes of different types of environmental managers.

Environmental managers and attitudes

To a certain extent, increasing uncertainty in EM can be linked to a shift in societal attitudes towards the environment from preindustrial conservation-orientated to industrial utilitarian attitudes. However, uncertainty also arises from the fact that different types of environmental managers are more or less conservationist or utilitarian in their environmental attitudes. This kaleidoscope of attitudes interacts in complex ways.

Table 4.1 represents a broad summary of the environmental attitudes of different types of environmental managers. There is potentially much variety in environmental attitudes within the different types of environmental managers, but it is still useful to treat these environmental managers in terms of general attitudinal groups.

The different types of environmental managers fall into one of three broadly

Table 4.1 Environmental managers and attitudes.

Environmental managers	Predominant environmental attitude	Associated EM practices
State	UTILITARIAN Some state agencies more utilitarian than others (e.g. fisheries vs environment). Possible shift in late 20th century towards more conservationist attitude overall.	Promotion of maximum fish catches, mineral extraction, etc.
Environmental NGOs	CONSERVATIONIST Some environmental NGOs are more conservationist than others (e.g. Earth First! vs Worldwide Fund for Nature), but all prioritize environmental conservation over exploitation.	Debt-for-nature swaps, prevention of whale killing, etc.
TNCs	UTILITARIAN Strongly utilitarian due to quest for maximum profits. Growing public pressure to adopt a more conservationist attitude.	Clearfelling of tropical and temperate forests, drift net fishing, etc.
International financial institutions	UTILITARIAN Strongly utilitarian reflecting the influence of EMDC state backers and market principles on which international financial institutions act. Under increasing public pressure to adopt conservationist stance.	Loans associated with mega-development projects (e.g. large dams, trans-migration), structural adjustment, etc.
Farmers and fishers	UTILITARIAN/CONSERVATIONIST Great diversity of attitudes depending on various factors (e.g. size of operations, link to global markets). Significant proportion of farmers and fishers guided by conservationist attitudes, even though utilitarian attitudes tend to predominate.	Industrial agriculture and extensive land clearance for cash crops, depletion of local fish stocks; preservation of remnant natural habitats on farms, restrictions on over-fishing, etc.
Nomadic pastoralists and shifting cultivators	CONSERVATIONIST Do not always adhere fully to conservationist attitudes, but the latter often required due to harsh environments that nomadic pastoralists and shifting cultivators manage. Increasingly forced to abandon conservationist attitudes due to restrictions on their EM practices.	Rotational agro-pastoral and agricultural practices with adequate regeneration of vegetation during fallow periods, use of organic fertilizers (e.g. ash, compost), etc.
Hunter–gatherers	CONSERVATIONIST Long-term, low impact EM practices generally based on spiritually-sanctioned conservationist attitudes. Some elements of utilitarian attitudes, but harsh environment often encourages the primacy of conservationist beliefs.	Selective culling and propagation of forest species, taboos against environmental degradation, etc.

distinctive attitudinal categories. Not surprisingly, given the discussion above, most environmental managers in recent times have been characterized by predominantly utilitarian environmental attitudes. Yet, the strength of the attachment of different environmental managers to utilitarian attitudes varies considerably. Most TNCs, international financial institutions and states can be categorized as possessing fairly strong utilitarian attitudes. There is certainly some variability even within these types. Within states, for example, different agencies reflect differing degrees of utilitarianism, depending on their formal EM responsibilities. Selected agencies within the state (e.g. environment agencies) may even adhere to moderately conservationist attitudes in line with their environmental "stewardship" functions. These types of environmental managers have, nonetheless, long reflected a predominantly utilitarian outlook on EM.

A second group of environmental managers base their EM policies and practices on broadly conservationist attitudes. On the one hand, there are the grassroots environmental managers who have long been guided by such attitudes. Hunter–gatherers, nomadic pastoralists and shifting cultivators have often based their activities on the need to manage their environment in a manner consistent with conservationist ideas. As this book has noted, these types of environmental managers have also sporadically pursued EM policies and practices more in keeping with utilitarian, rather than conservationist, attitudes (see Ch. 3). Yet, these types of environmental managers – influenced by the fact that they often need to manage harsh environments – have been guided most frequently by conservation-orientated attitudes (Hong 1987, Ingold et al. 1988). On the other hand, and as will be discussed more fully in the case study below, environmental NGOs are a relatively new type of environmental manager that has developed in recent decades, precisely in order to encourage EM policies and practices based on conservationist attitudes (Bowlby & Lowe 1992).

Between these two broad camps, a third category encompasses farmers and fishers – in aggregate, the most numerous environmental managers. As noted below with reference to farmers, there is perhaps a greater diversity of attitudes – ranging from the highly utilitarian to the staunchly conservationist – among these than with any other type of environmental manager (Carr & Tait 1991, Wilson 1996). The influence of utilitarian attitudes has often been strongest and has been reflected in intensive industrial farming and fishing practices, which have made a major contribution to environmental degradation (Briggs & Courtney 1989, Fairlie 1995). The following case study of farmers' attitudes towards the environment concentrates on those whose attitudes are at the utilitarian end of the spectrum.

Among farmers, it has been those who have needed to manage new lands who have displayed utilitarian environmental attitudes most clearly. Such "pioneer" farmers have also been a major factor in environmental degradation, especially in recent centuries, thereby contributing to increasing uncertainty in EM (Williams 1989). Wynn (1979: 171) has suggested that pioneer farmers usually "show little concern for the aesthetic or long-term consequences of their action and, on the whole, they value immediate and concrete benefits more highly than such intangible concerns as resource conservation".

Three factors, in particular, are associated with the utilitarian attitudes of pioneer farmers. First, they typically believe natural resources in their new homelands to be inexhaustible. This was particularly evident in European colonial expansion between the fifteenth and early twentieth centuries (Wynn 1977, 1979, Short 1991), when Europeans invaded lands that had hitherto been subject to EM policies and practices of indigenous farmers, shifting cultivators or hunter–gatherers that, by and large, were based on more conservation-orientated environmental attitudes (e.g. North America, Australia). Although the environment that Europeans encountered was not "untouched", the pioneer farmers often perceived it to be in a pristine state (Nash 1967, Williams 1989). This perception was important because to these farmers their new environment appeared limitless in extent and its natural resources seemed inexhaustible. Not surprisingly, under these conditions utilitarian environmental attitudes thrived.

In the North American context, Turner (1920) described this process with reference to the "frontier thesis". The frontier is an imaginary boundary beyond which limitless environmental resources seem to stretch. As long as this frontier is perceived to exist, it encourages a utilitarian attitude towards the environment (Wynn 1979; see also Ch. 8). This process was especially noticeable during the colonial era, when forests in many parts of the world were destroyed by pioneer farmers driven by such attitudes. Although the effects of the utilitarian attitudes of pioneer farmers were felt in most parts of the colonial world at one time or another, perhaps the most dramatic example relates to "New World" forests. In the USA, Canada, Australia, New Zealand and South Africa, for example, hundreds of thousands of square kilometres of forest were destroyed between the late eighteenth and early twentieth centuries, partly because of deliberate forest destruction by pioneer farmers seeking to produce food and fibre for a globalizing capitalist market (Tucker & Richards 1983, Fleet 1984, Williams 1990, Grove 1993). A similar process can be seen today in various parts of the world, notably in Brazil's Amazonia, where pioneer farmers – often swayed by utilitarian attitudes – are in the process of clearing vast and seemingly "inexhaustible" tracts of tropical forest (Hecht & Cockburn 1989).

The origins of pioneer farmers can also throw light on why these environmental managers tend to hold utilitarian environmental attitudes. Pioneer farmers of European origin, for example, left a familiar tamed environment for an often unknown New World wilderness (Nash 1967, Shepard 1969, Short 1991). Upon arrival in their new homelands, these farmers typically attempted to convert this "wilderness" (often carefully managed by indigenous environmental managers) into a more "congenial" environment – that is, an environment that resembled their place of origin more closely. Since the vast majority of these farmers hailed from areas of Europe long characterized by highly modified environments, the clearance of New World wilderness was an essential part of rendering their new environment "habitable".

Yet, the utilitarian attitudes of pioneer farmers were not solely reducible to the quest for economic advancement. Thomas (1956) and Wilson (1992b), for example, have suggested in the Mexican and New Zealand context that farmers not only cleared forests for economic reasons (timber, agriculture) but also sought to re-create the treeless landscapes of Castile (Spain) and Scotland (Great Britain) in their new homelands. Their new home became more "beautiful" as the forests were removed. British pioneer farmers, in particular, were anxious to transform wilderness into a pastoral idyll based on highly utilitarian environmental attitudes where "nature in the raw" was perceived as menacing, threatening, alien "wastelands" (Shepard 1969, Wynn 1979). A central task of New World farmers was the transformation of these alien environments into "cultural" landscapes (Nash 1967, Williams 1989, Wilson 1992b). These environmental managers were reinforced in this quest by the prevailing technocentric worldview (itself strongly influenced by Judeo-Christian notions that elevated humans above nature) that encouraged the development of "productive" landscapes.

A final factor influencing the development of utilitarian attitudes among many pioneer farmers relates to the problems that they encountered in adapting to an "alien" environment. Fear of the unknown helps in part to explain the fervour with which pioneer farmers sought to manage the environment in line with highly utilitarian attitudes. In New Zealand, Australia and the USA, for example, early pioneers felt "overwhelmed", "scared" and "engulfed" by the unfamiliar forest (Frawley 1987, Williams 1989, Wilson 1992b). How this fear of the unknown reinforced utilitarian attitudes among many pioneer farmers may be contrasted with how a similar environment prompted many hunter–gatherers to adopt conservation-orientated attitudes (see above). The latter's response often reflected a spiritually based fear of alienating the gods that were believed to be living in the environment, whereas the former's was driven by the fear of "dangers" that lurked in the forest, amplified in turn by religious notions of "mastery" over the environment.

This discussion of pioneer farmers has explored some of the EM implications of environmental attitudes based on utilitarian ideas. Certainly, not all farmers have adopted utilitarian attitudes similar to these pioneers. However, although many farmers may favour EM policies and practices based on conservation-orientated attitudes, the growing dependency of most farmers on the global capitalist economy, and the associated reliance on industrial-type agricultural techniques, suggests that, on balance, a majority of today's farmers have to be utilitarian in outlook (McDowell & Sparks 1989, Wilson 1996). This trend is increasingly challenged by the growing influence of conservationist attitudes in society. Already today, evidence from Europe and North America suggests a possible shift in the environmental attitudes of some farmers from utilitarian to conservation-orientated attitudes (Kreutzwiser & Pietraszko 1986, Carr & Tait 1991, Morris & Potter 1995, Wilson 1996). To appreciate the importance of this countervailing influence fully, we now turn to an assessment of the attitudes of managers in environmental NGOs.

Table 4.1 (above) illustrated that utilitarian ideas have, on balance, been more influential in shaping the environmental attitudes of different types of environmental managers. Growing popular disenchantment with the environmental degradation that is often a practical outcome of the adoption of utilitarian attitudes has prompted the increasing popularity in society as a whole of conservationist attitudes (O'Riordan 1995b, McCormick 1995, Dobson 1995). The possible connection between this potential shift in environmental attitudes and multi-layered EM is twofold. On the one hand, it may be associated with a comparable shift in the environmental attitudes of selected environmental managers – political leaders, state bureaucrats, TNCs, or international financial institutions, for example. On the other hand, it may serve to enhance the position of those types of environmental managers especially noted for their conservationist attitudes. In this regard, the role of environmental NGOs as potential societal "opinion-formers" on environmental matters may be of crucial importance. It is, therefore, important to be clear about the meaning that environmental NGOs attach to conservationist ideas.

Not only are environmental NGOs one of the most recent arrivals in multi-layered EM, they are also the only environmental manager whose main objective is to change the attitudes of other environmental managers and the public in line with conservationist ideas (see Table 2.2). There are, of course, different "shades of green" in the attitudes of environmental NGOs (see below). However, and as with the preceding discussion of pioneer farmers, it is useful here to explore the diverse motivations that have engendered the conservationist attitudes of these environmental managers.

First, and as noted above, the revulsion of the urban middle classes in nineteenth century Europe and North America against the social and ecological effects of industrial development was associated with the development of a "conservation movement". The so-called "romantic" thinkers (e.g. Thoreau [USA], Wordsworth [Great Britain], Eichendorff [Germany]) were part of a major attitudinal change that was central to the spread of conservationist attitudes. They were critical of industrial "progress" and the associated social and environmental uncertainty linked to the ensuing degradation of the environment (Pepper 1984, Brimblecombe 1987). The growing disenchantment with industrial and increasingly urban development prompted a glorification of the "natural" environment that was, in turn, equated with non-urban areas (i.e. countryside, mountains). This anti-urban and anti-industrial mood was an important contributing factor in the growth of early conservationist initiatives aimed at preserving selected natural habitats (i.e. national parks), and it formed the basis for early environmental groups such as the Sierra Club in the USA (established 1892), and the National Trust in Great Britain (established 1894) (Short 1991, Bowlby & Lowe 1992, Micklewright 1993).

Secondly, and complementing this reaction against industrialization, was growing concern over the rapid depletion of forests and other natural resources, and the associated awareness that such EM practices were leading to resource scarcity and environmental degradation. An early critic of these practices, and the associated utilitarian attitudes upon which they were typically based, was George Perkins Marsh. He was particularly incensed by deforestation worldwide: "we have now felled forest enough everywhere . . . Let us restore this one element of material life to its normal proportions, and devise means for maintaining the permanence of its relations to the fields, the meadows, and the pastures" (Marsh 1965: 280). As one of the founders of the contemporary conservation movement in the Anglo-Saxon world, Marsh's ideas were especially influential in shaping the content of conservationist attitudes (Pepper 1984). In the twentieth century, a growing emphasis on the idea of absolute limits to possible resource exploitation further sharpened concerns over resource scarcity. That emphasis was notably reflected in the work of neo-Malthusian thinkers who argued in the 1960s and 1970s that runaway resource exploitation based on utilitarian attitudes was resulting in a situation whereby human EM practices threatened to breach absolute limits for such exploitation (Meadows et al. 1972). By this time, the conservation movement was gathering political and social momentum and environmental NGOs were a notable institutional manifestation of this trend (McCormick 1995).

Thirdly, the conservationist attitude of environmental NGOs also partly derives from a strong moral sensibility that it is wrong for humans to eradicate other species. Although there is some recognition here that flora need to be protected from unsustainable EM practices, this moral sensibility is undoubtedly strongest with reference to fauna (Benton 1993). Many of the oldest environmental NGOs were created in order to lobby for the protection of endangered species – the Society for the Protection of Birds (established 1881 in Great Britain) and the Natal Game Preservation Society (established in 1883 in South Africa), are good examples – and this concern continues to be reflected in the activities of many modern environmental NGOs such as Greenpeace (e.g. the campaign against whaling) and Friends of the Earth (e.g. the campaign on behalf of the black lemur of Madagascar). The common concern linking together these various campaigns by environmental NGOs is based on the notion that "non-human beings have intrinsic value" (O'Neill 1993: 9–10), and that such a recognition must be embodied in EM.

Environmental NGOs pursue environmental policies and practices based on conservationist attitudes linked to anti-urban, anti-industrial, pro-wilderness and species-protection concerns. As with farmers, and for that matter all other types of environmental managers (see Table 4.1), there are notable differences in attitudes within the environmental NGO community. However, although the range of environmental attitudes among farmers, for example, may range from the fiercely utilitarian to the conservationist (Morris & Potter 1995, Wilson 1996), the comparable range among staff of environmental NGOs is much smaller and is confined to the conservationist end of the spectrum. Environmental NGOs are, nonetheless, differentiated according to "different shades of green", reflecting the depth of their attachment to ecocentric worldviews. The attitudes of some managers in environmental NGOs reflect a "shallow approach" (i.e. reformist) to conservation, whereas the attitudes of others is indicative of a "deep approach" to the issue (McCormick 1995, Dobson 1995). Although the reasons for these differences remain largely unexplored, it would appear reasonable to expect a strong correlation between socio-economic status and the extent of "greenness" (Lowe & Goyder 1983, Eden 1993, Cowell & Jehlicka 1995).

This discussion has explored the environmental attitudes of different types of environmental managers in order to highlight the sociocultural dimensions to uncertainty in EM. However, comments with reference to environmental NGOs and farmers indicate further complexity associated with the environmental attitudes of individual (as opposed to group) environmental managers. The adoption of these attitudes is linked to diverse social and economic factors, including education, income, urban/rural residency, and gender (Inglehart 1977, VanLiere & Dunlap 1980, Kellert 1984, Choi 1985, McAllister 1994).

Education, for example, is frequently seen as a critical factor in the determination of environmental attitudes of individual environmental managers. Specifically, environmental managers with higher formal education tend to be more aware of the implications of environmentally destructive behaviour than their poorly educated counterparts. This is seen to be one of the bases for their greater support for conservationist measures (McDowell & Sparks 1989, McAllister 1994). In contrast, the possible role of gender in the shaping of environmental attitudes of individual environmental managers is emphasized increasingly (Jackson 1994, Warren 1994). The argument is made that, whereas female environmental managers may tend to adopt more conservation-orientated attitudes, their male counterparts more frequently adhere to utilitarian attitudes. Various reasons have been suggested for this discrepancy, notably male domination of political, economic and scientific activity Merchant 1982), or even women's inherent "closeness to nature" (Shiva 1989).

These and other socio-economic characteristics relate to attitude formation in complex ways. What is clear, though, is that these relationships add to the uncertainty surrounding the cultural influences on EM policies and practices. Such uncertainty is exacerbated because of the necessary interaction of the environmental worldviews and attitudes of different environmental managers, an issue explored in the next section in the context of a discussion of environmental discourses of different types of environmental managers.

Environmental discourses

Environmental discourses link together the environmental attitudes and worldviews of different environmental managers. They present an "argument" for given sets of EM attitudes and practices, as well as a social and ecological justification for those attitudes and practices. These discourses are "frameworks that embrace particular combinations of narratives, concepts, ideologies and signifying practices", each of which is relevant to a particular area of environmental and social action (Barnes & Duncan 1992: 8). Different environmental managers may promote different environmental discourses, with the result that multiple discourses are a prominent cultural attribute of multi-layered EM, adding potentially further sociocultural uncertainty to such EM.

Environmental discourses may be related to highly specific EM issues and practices, or may be associated with broad themes of human–environment interaction (Myerson & Rydin 1996). Further, different environmental managers may promote different environmental discourses, although this need not always be the case. As noted below, different types of environmental managers often subscribe to the same environmental discourse as part of the promotion of congruent interests and attitudes. Indeed, the purpose of such discourses is precisely to try and create a common understanding of environmental issues and their management implications. Such an endeavour is proving increasingly difficult in the context of intensifying conflict in multi-layered EM, rendering the quest for predictability ever more complicated. Two examples of environmental discourses serve to illustrate this point.

The first relates to the growth of a discourse of scientific forestry linked to a highly specific understanding of the utility and purpose of selected forests. This discourse developed in the nineteenth century, but is still prevalent in many parts of the world today. It suggests that the utility of forests resides primarily in the promotion of long-term commercial exploitation of selected key species for use in the global capitalist market. Central to this discourse was the quest to measure "progress" through quantitative indicators: volumes of timber extraction, export figures, rate of replanting, or financial revenue (Bryant 1996).

This discourse inevitably privileged certain environmental managers over others. As a form of EM, scientific forestry has been a vehicle for promoting the interests of states and businesses who possess the legal, political, or financial means to pursue this EM practice (Guha 1989, Peluso 1992). However, those sectarian interests have been represented through environmental discourse as reflecting interests in the "common good". Scientific forestry is portrayed as an "ecologically sound" EM practice that also promotes social wellbeing through the generation of revenue and employment.

An integral part of the discourse of scientific forestry has been an associated attempt to belittle the EM policies and practices of environmental managers with competing claims on the forests. Hunter–gatherers or shifting cultivators, for example, have been portrayed as not only the enemies of scientific forestry but also as opponents of "sustainable" EM policies and practices (Jewitt 1995). Indeed, these traditional environmental managers only received a favourable assessment when they were incorporated in practices associated with scientific forestry. Karen shifting cultivators in colonial Burma were considered "evil" by forest officials until a role in reforestation was found for those cultivators, at which point they became "good" foresters (Peluso 1992, Bryant 1994b). In this manner, the discourse of scientific forestry not only specifies "good" or "bad" EM practices, but, by extension, also differentiates between types of environmental managers based on those practices.

The second example concerns the environmental discourse on "sustainable development". This discourse is about the attempted reconciliation of continued economic growth and environmental conservation, and has become increasingly influential in national and international policy-making circles. Once again, here at a general level, assertions about "correct" forms of human–environment interaction are made, based on the general idea that the global capitalist system needs to be perpetuated (Redclift 1987, Reid 1995). In the process, other ideas – for example, linked to a perceived need for revolutionary change and the abandonment of that system – are not considered (Sachs 1993).

Not only does the discourse of sustainable development favour capitalist over non-capitalist practices, it also tends to favour those environmental managers who benefit disproportionately from the global capitalist system, namely, states, TNCs and international financial institutions (Ekins 1993, Silva 1994). In contrast, those environmental managers who are the primary losers under that system (e.g. poor farmers, shifting cultivators, fishers) not surprisingly have little opportunity to shape the discourse on sustainable development. Their "participation" and "approval" may be solicited by traditionally powerful environmental managers seeking to implement sustainable EM policies and practices, but the role of marginal actors in multi-layered EM remains firmly subordinate to more powerful environmental managers – and this hierarchy is implicitly sanctioned through the environmental discourse of sustainable development (Lele 1991).

Discourses on scientific forestry and sustainable development have not succeeded in reconciling the differing interests and attitudes of environmental managers. Therefore, they have not helped to reduce uncertainty in EM. On the contrary, these and other environmental discourses have generated massive opposition from environmental managers threatened with further marginalization because of the specific EM practices promoted through those discourses.[1] Community forestry, for example, has been promoted to counterbalance scientific forestry. Its proponents suggest that the former is much better placed than the latter to deliver both social justice and environmental conservation in terms of forest management (Poffenberger 1990, Jewitt 1995). Just as with the discourse of scientific forestry that it seeks to displace, the discourse of community forestry privileges certain environmental managers (e.g. shifting cultivators, small-scale farmers) over other environmental managers (e.g. logging companies). Similarly, but at a more general level, an "anti-development discourse" has emerged in recent years to challenge directly the discourse of sustainable development and its assertion of the continued viability of the

1. The material consequences of such opposition are considered in Chapter 5.

global capitalist system (Watts 1993). The argument here is that the world's growing social and environmental problems can never be resolved through development under the global capitalist system precisely because that system is the root cause of these problems (Sachs 1993). Instead, this discourse asserts the need for a more locally based and non-capitalistic set of EM policies and practices (Pepper 1993).

Environmental discourses are therefore the subject of growing dispute among environmental managers, not least because these discourses favour certain interests and activities over others. Rather than serving to reconcile the often different environmental worldviews and attitudes of environmental managers, environmental discourses serve mainly as a means to assert and justify those differing worldviews and attitudes and the EM interests associated with them. Thus, the existence of multiple environmental discourses has only added to uncertainty in EM.

Conclusion

This chapter has illustrated the complex relationship between culture and multi-layered EM through its discussion of environmental worldviews, attitudes and discourses of environmental managers. The end result is to emphasize once more the growing uncertainty that characterizes the process of EM.

Environmental managers may adhere to conservationist or utilitarian environmental attitudes, situated within the broader context of contrasting ecocentric or technocentric worldviews. Generally speaking, environmental managers who hold conservationist attitudes also generally subscribe to an ecocentric worldview, whereas those who adhere to utilitarian attitudes often hold technocentric worldviews. This relationship is often more complicated than this summary would suggest, in that specific attitudes are conditioned by a range of socio-economic characteristics that belie easy description. The result is multi-layered EM in which environmental managers – often holding different environmental worldviews and attitudes – must coexist. Such a situation is a source of much uncertainty in EM. Yet, that uncertainty is compounded because of the efforts of different environmental managers to assert and justify their individual worldviews, attitudes and EM practices in relation to other environmental managers. The resulting discursive complexity only adds to the uncertainties of the process of EM.

Overall, Part II of this book has suggested a scenario in which EM has been characterized by growing environmental and social uncertainty. Chapter 3 pursued this theme in relation to the intensifying material impact of environmental managers on the environment, whereas Chapter 4 has related that material impact to cultural issues that in and of themselves have usually exacerbated that uncertainty. It is important, nonetheless, to appreciate the ways in which environmental managers seek to combat uncertainty through the pursuit of predictability in EM. The latter needs to be related to political, market and policy considerations, the subject of Part III of this book. The endeavour there will be to consider whether developments in these spheres hold the prospect for a reduction in the uncertainty that is central to the operations of multi-layered EM.

PART III

TOWARDS PREDICTABILITY?

Part II has highlighted how EM practices and cultural perceptions – interacting in a complex and often contradictory manner – have contributed to increased intensity of human use of the environment. This has led, in turn, to pervasive environmental degradation and, more generally, increased uncertainty in EM.

The response of state and non-state environmental managers to uncertainty in EM has been a quest for predictability. This point was illustrated analytically in Chapter 2 with reference to the various dimensions of the concept of predictability. To understand how environmental managers have sought predictability in practice is to relate their activities to a broader context. It is necessary to place this quest in a social and environmental context conditioned by political, market and policy-related considerations in multi-layered EM.

As Chapter 5 shows, EM is an integral part of the political process in that decision-making is linked to unequal power relations in the allocation and use of environmental resources. Just as EM can be thought of as a multi-layered process, so too the link between EM and politics is more complex than is usually assumed. As Chapter 3 highlighted, increased uncertainty in EM has been related to the development of the global capitalist system. This theme is pursued in Chapter 6 where the role of the market in relation to multi-layered EM is explored. In contrast, Chapter 7 takes up the link between policy and EM to explore how environmental managers formulate policies that may serve as the basis for sustainable EM practices.

The theme that integrates these three chapters is the role that political, market and policy considerations play in the quest by environmental managers for predictability in EM. Whether those considerations enhance or hinder that quest is also a theme running through Part III of the book.

CHAPTER 5
Environmental management and politics

A key feature of a re-evaluation of EM must be the recognition that it is a politicized process. Human–environment interaction is imbued with political meaning, and EM is embedded in politics. This book emphasizes the role of power relations and conflict between environmental managers. A central task in this chapter is to relate the understanding of EM as a multi-layered process to an appreciation of how power relations condition EM practices, and how they may affect different environmental managers within that process. At issue here is how power relations influence the ability of environmental managers to enhance predictability in EM.

Power and multi-layered environmental management

To understand the interactions of environmental managers operating within multi-layered EM, it is important to explore the ways in which unequal power relations figure centrally in their policies and practices. The power of the state must not be exaggerated and the importance of non-state environmental managers must not be understated – notably peasants, hunter–gatherers and other grassroots environmental managers, who are typically viewed as weak and powerless. As Giddens (1979: 149) reminds us, "all power relations, or relations of autonomy and dependence, are reciprocal: however wide the asymmetrical distribution of resources involved, all power relations manifest autonomy and dependence 'in both directions'". In relating EM to politics, it is therefore essential that the complexities inherent in power relations be fully acknowledged.

It is useful to understand power relations in EM in relation to the concepts of uncertainty and predictability developed in Chapter 2. There are three aspects that need to be considered here. All environmental managers seek predictability in their EM practices, but their political and economic status in society means that they are not all equally positioned to pursue that quest. For example, state officials often command financial, technological and human resources on a scale well beyond that enjoyed by non-state environmental managers such as fishers or hunter–gatherers. This point is perhaps most clearly illustrated with reference to technological resources. State officials often enjoy privileged access to state-of-the-art technologies that facilitate exploration, extraction, transport and processing of environmental resources. These technologies may also include the latest computer-based facilities linked to the creation of complex models or "virtual realities" (see Ch. 8). Unequal access to these resources conditions the ability of environmental managers to pursue predictability in EM. It should not be automatically assumed that state environmental managers are always better positioned than non-state environmental managers in this regard. The power of state officials may be limited because of the relative strength or weakness of the institution for which they work. In some ELDCs, for example, the power of state officials to enhance predictability may be limited in practice. In Senegal, for example, the interpenetration of business and state interests is such that formal environmental regulations are highly ineffectual (Ribot 1993). Conversely, powerful non-state environmental managers, such as TNCs and wealthy farmers, may possess power that rivals that of many state officials in a given country. In the USA, for example, farmers' lobbies have put pressure on Congress and the Department of Agriculture to maintain agricultural prices at artificially high levels, thereby increasing financial returns for agricultural commodities (Browne et al. 1992). Indeed, "farm-producer and farm-industry groups [have] long maintained close relationships with the USA Department of Agriculture and [have] been able to exercise effective control over its policies" (Hays 1987: 295). The manifestation of power relations in EM is associated, therefore, with an unequal ability to pursue predictability in EM.

All environmental managers face environmental uncertainty. As discussed with reference to Beck's (1992) "risk society", in some cases of environmental degradation uncertainty may be equally distributed among members of society (e.g. nuclear fallout, as from the Chernobyl incident), but in others uncertainty is unequally distributed (Blaikie & Brookfield 1987, Bryant 1992). Unequal power relations are manifested not only in the ability to pursue predictability in EM, but also in the degree of exposure of environmental managers to various types of uncertainty in the first place. For example, hunter–gatherers may be less able to respond to environmental degradation (e.g. loss of forest) than are other environmental managers, such as state officials or decision-makers in TNCs, international financial institutions, and environmental NGOs. Hunter–gatherers rely on little-disturbed environments for survival, whereas TNCs, for example, are often able to move away from areas that have been severely degraded. This is not simply a question of proximity to the site of environmental degradation, since unequal exposure of on-site environmental managers is often differentiated according to wealth. Typically, poorer environmental managers are more susceptible to environmental degradation than are their wealthier counterparts; soil erosion, for example, will hit poor farmers on ecologically marginal land particularly hard (Blaikie 1985, Thapa & Weber 1991). Further, unequal exposure may also be linked to knowledge about environmental degradation; for example, well informed environmental managers may be in a better position to react to certain types of environmental degradation (e.g. toxic spills) than their less informed counterparts (Beck 1992). Wealth and knowledge often are interlinked so that the more powerful in society are usually best able to reduce uncertainty for themselves. Power relations condition to some extent at least the ability to reduce exposure to uncertainty (Boehmer-Christiansen 1994a).

Such is the link between the existence of uncertainty and the promotion of predictability that the ability of some environmental managers to enhance predictability may lead directly to increased uncertainty for other environmental managers. From this perspective, power relations in EM are partly about the ability to transfer uncertainty from one environmental manager to another. An example here is how state officials designate specific areas for permanent protection (e.g. national parks), but in so doing may displace other environmental managers, forcing them to migrate to less suitable locations. The displacement of the Maasai to make way for national parks in Kenya is a case in point. The creation in the early twentieth century of state wildlife reserves in areas traditionally used by Maasai nomadic pastoralists led to the latter's exclusion from the parks. Such displacement resulted in growing resistance by these environmental managers, as manifested in the killing of large game in these reserves (Peluso 1993). In this instance, the quest for environmental predictability by one environmental manager (i.e. state officials in charge of game preservation) results in increased political and environmental uncertainty for another (i.e. loss of rangeland, coercion).

The politics of EM is, therefore, not only about unequal exposure to uncertainty and the unequal ability to enhance predictability in EM, but is also about the power to transfer uncertainty from stronger to weaker environmental managers. This is not to say that power relations in EM are necessarily a zero sum game; that is to say, that the quest for predictability by some environmental managers must be at the expense of the ability of others to pursue a similar objective. It is conceivable that the quest for increased predictability in EM by one type of environmental manager – and here the state springs readily to mind – may lead to greater predictability for most, if not all, environmental managers (and environmental users). Efforts by states to combat the depletion of ozone are a case in point. Here, successful reduction in environmental uncertainty, associated with the adverse health implications of stratospheric ozone depletion linked primarily to chlorofluorocarbon emissions, will benefit all (Benedick 1991, Rowlands 1995).

Most EM issues do not lend themselves to such a common quest for predictability. Rather, most environmental degradation issues bear directly upon unequal power relations in EM. Hence, and as the above example of the Maasai highlights, EM is frequently a political process in which the power to shift uncertainty plays a central part. To appreciate this situation fully, it is necessary to examine in more detail the various environmental managers that may be involved, and how their EM activities may be conditioned by those of other environmental managers.

Political aims and interests

The interaction of environmental managers in the course of pursuing predictability is inevitably linked to political processes. In this regard, it is important to appreciate the political aims and interests of environmental managers. The last two sections of this chapter address how such aims and interests are associated with both conflict and cooperation among environmental managers. Here, the objective is to examine those aims and interests that motivate diverse environmental managers.

State environmental managers

As noted in Chapter 3, a central feature of the development of multi-layered EM was the emergence of the state as an environmental manager with increasing power over non-state managers. As this chapter notes, nonetheless, it may now be that the power of state environmental managers is generally on the wane, as other environmental managers, notably TNCs, international financial institutions and environmental NGOs, assume growing importance in EM. Yet, even today state environmental managers often play a pivotal role in EM decision-making (Hurrell 1994).

Although the specific ways in which state EM has manifested itself vary from country to country, a common theme is the quest for centralized political control over peoples and environments within the territorial jurisdiction of the state. Such control has been linked to the pursuit of commercial objectives in the global capitalist system, often but not always in conjunction with private business interests (see below). Yet, as Chapter 3 noted, the interests of state environmental managers cannot be reduced to those of the business groups with whom they may be closely associated.

It is important to emphasize that state EM is not a generic process or activity – that is, "state EM" is a general category that encompasses a diversity of specific environmental managers and EM practices. Further, state EM is conditioned by the formal political process insofar as constitutional arrangements and electoral practices influence the possibilities and constraints of state environmental managers. How states relate to each other also has a bearing on state EM practices (see Ch. 7).

As states have developed, they have become internally more complex, with management tasks allocated institutionally according to functional criteria. Since the emergence of the modern state in the eighteenth century, how the environment has been managed by the state has become centred on the activities of specialist departments and services such as forest departments, agriculture departments, or mining departments. This process of functional definition occurred notably in the eighteenth century in Europe and Japan, and in the nineteenth and early twentieth century elsewhere (Heske 1938, Tucker & Richards 1983, Totman 1989).

As a result of such specialization, these departments developed a particular set of institutional interests and objectives, and even a particular set of educational prerequisites for entry to service. The result was that knowledge about the specific environmental resource increased, but only at the expense of a broader integrated appreciation of EM. Forest departments were established with a brief to manage forests in keeping with maximum sustained yield and commercial exploitation, and were staffed by specially trained officials versed in the complexities of scientific forestry (Bryant 1997). A similar process of institutional specialization and elaboration occurred in other departments such as agriculture and mining. Within these departments, trained staff have not only focused on a particular resource, but they have also been moved, in their normal career development, from assignment to assignment as part of a broader set of management objectives. Indeed, as Furnivall (1956: 77) observed with reference to colonial Burma and Indonesia, "none of these [specialist] officials saw life whole and, by reason of frequent transfers, none of them saw it steadily". As noted below, one implication of such functionally defined state EM is the development of bureaucratic conflict, as different departments clash over contradictory management objectives (Pretty 1995).

State EM is further influenced by the way in which states are structured. The way that powers are allocated and defined in national constitutions may have a direct bearing on the nature and extent of state EM. For example, states are organized on the basis of the relative centralization or decentralization of political power among their constituent political units. At one extreme is the unitary state, where power is concentrated in the hands of a national government, and where there is a notable lack of countervailing power at the local or regional levels. In such a context, state EM is essentially synonymous with EM practised by officials working for a centralized authority. In countries such as the UK, France, Burma and Thailand, state EM conforms largely to this pattern. In contemporary Burma (Myanmar), EM is exclusively the formal preserve of the centralized Burmese state, and specifically the forest and other natural resource departments (Bryant 1997). Similarly, in the UK, although local authorities have limited EM powers, EM is predominantly the preserve of the national government (Lowe & Goyder 1983, Robinson 1992, Garner 1996).

In other cases, the political structure is characterized by a constitutional division of powers between different levels, usually between a central or federal government and provincial or regional governments. In such "federal" systems, a major source of political dispute has been over the allocation of EM tasks and responsibilities (and revenue), with predominant control in this area residing typically with one or the other level of government. For example, in the USA predominant control over state EM resides with the federal government, whether it be in areas of forest management, offshore oil drilling, soil conservation or pollution control (Lester 1990). Indeed, the trend in recent decades has been seemingly for successive federal administrations in the USA to limit "the ability of states to engage in environmental government, by reducing grants to enhance their management capabilities" (Hays 1987: 536). Individual states in the USA nonetheless are often responsible for implementing federal legislation and they control such areas as emission thresholds and waste disposal and may be acquiring, on balance, additional powers in diverse environmental sectors (Hays 1987, Miller 1994).

In other countries, control over EM is located more at the provincial or regional level, particularly EM activities associated with natural resource exploitation. For example, in countries as diverse as Canada, Germany and Malaysia, control over state EM practices linked to natural resource exploitation and conservation has largely been the preserve of the provincial or regional government. In Malaysia, control over forest management resides with the provincial governments; hence, controversy over excessive logging in Sarawak is linked largely to the unsustainable EM policies and practices of the Sarawak government (Hong 1987). Similarly, in Canada many aspects of EM reflect overlapping or concurrent jurisdiction under the Canadian constitution, yet provincial governments derive much of their power as environmental managers from their ownership of publicly owned lands and the natural resources associated with those lands (Mitchell 1995). Thus, "through its ownership of public lands and its legal authority to dictate how its natural resources . . . will be managed, developed, and/or conserved, a provincial government is pivotal in affecting the quality of the environment within its borders" (Skogstad & Kopas 1992: 45). The situation in Germany follows along roughly similar lines, in that control over natural resource exploitation (notably forestry) is largely the preserve of the provincial Länder (Jones 1994). Thus, in countries characterized by federal constitutions, state EM is not necessarily synonymous with EM by the national government. Rather, it is fragmented among various levels of government, with the specific allocation of EM powers reflecting historical and political developments in the country in question. Indeed, as the preceding discussion highlights, provincial governments may have greater overall responsibility for managing the environment than their national counterpart.

Further complexity surrounding the constitutional allocation of EM responsibilities has been associated with the emergence of new EM problems, typically associated with adding to the environment (see Table 2.1). Problems associated with toxic waste disposal or nuclear energy – issues not even remotely conceivable at the time that many constitutions were drawn up – have been the subject of complex political and legal negotiations between different levels of government (let alone inter-state debates; see below). Such negotiations have been different, in keeping with the various management implications of new environmental problems. Whereas different levels of government have tussled over natural resource management because of the revenue attached to controlling such management (e.g. timber), in the case of the new environmental problems (typically the by-product of industrial development) attention has centred on the problems of allocating responsibilities for who will regulate pollution clean-up (Hays 1987, Rees 1990, Weale 1992). Indeed, these continuing constitutional negotiations over state EM are one of the political manifestations of the "risk society", that is, the issue of which agency and level within the state is to assume ultimate responsibility for the management of increasingly risky by-products and practices of industrialized societies.

Beyond the question of the internal structure of the state itself, state EM is also affected by the way in which the state in general is related to civil society. Specifically, state EM can be conditioned by differing electoral systems that empower or reward certain environmental managers over other environmental managers through access to political power, and the political and economic benefits that often flow from such access. From an EM perspective, those benefits can be diverse. Perhaps most obviously, electoral success may be associated with favouritism towards certain types of environmental managers in the allocation of rights to exploit the environment – logging, mining, fisheries, and so on (Rush 1991, Marchak 1995). Perhaps the most vivid example of such favouritism (or "crony" capitalism) relates to EM policies and practices in the Philippines during the Marcos era. In the 1970s and early 1980s, the right to exploit national resources (e.g. timber, fish, minerals) was largely based on connections to President Ferdinand Marcos and his family rather than on other factors (Broad 1993, Vitug 1993).

Less evident is how access to political power via the electoral system may also be linked to greater environmental knowledge through the ability to tap information systems controlled by the state. Such privileged access to knowledge can empower non-state environmental managers, but it simultaneously affects the EM capacity of state environmental managers. The point here is not that the latter are unable to pursue their own institutional interests without the validation of non-state environmental managers; rather, it is to emphasize that state environmental managers typically are not in a position to pursue those interests without often taking into account the interests of non-state environmental managers. State EM is associated intimately with the interests and practices of non-state environmental managers.

How state EM is affected by the electoral system varies from country to country, depending on the specific structure of that system. At one extreme is an electoral system that is essentially designed to validate the EM policies and practices of the state. In countries in which authoritarian political systems are the norm (e.g. contemporary Burma, Cuba, North Korea, Libya), the opportunity for environmental managers outside the state to use the electoral system to challenge or modify state EM is severely limited or even non-existent (Bryant 1997). However, the majority of countries today operate under democratic, or quasi-democratic, political arrangements in which outside environmental managers have a greater or lesser ability to challenge or influence state EM through electoral politics. In the "established" democracies (e.g. USA, Australia, Germany), state environmental decision-making is under sustained and detailed public scrutiny, and this can be seen in the modification of state EM policies and practices (McAllister 1994). Increasingly today, a similar process is occurring in many ELDCs, notably in such countries as Thailand, Brazil, the Philippines and South Africa (e.g. Rush 1991, Hurrell 1992).

The relationship between democracy, electoral systems, state EM and non-state environmental managers is complex. Political systems characterized by a "first-past-the-post" system (i.e. a system in which the candidate with the majority of the votes within an electoral constituency wins the seat) disenfranchize marginal political groups – such as the Green Party in the UK – while they tend to favour the articulation of mainstream political and economic interests through the established political parties. Indeed, as Lowe & Goyder (1983: 73) emphasize, the "first-past-the-post electoral system . . . militates against small parties that are not geographically concentrated". Under this system, it is exceedingly difficult for non-mainstream groups to challenge state EM through the electoral system. Conversely, this system may enhance the power of capitalists, who can influence mainstream political parties (perennially starved for funds) competing within this electoral system in order to condition the nature of state EM indirectly. Efforts to prevent the adoption of rigorous EU pollution standards in the UK by powerful business interests (represented notably by the Confederation of British Industry) with links to the Conservative Party are a case in point (Weale 1992, Flynn & Lowe 1992). In contrast, political systems characterized by proportional representation enable marginal groups – such as Die Grünen (Green Party) in Germany – to obtain sufficient political power to influence the direction of state EM policies (often in coalition with other parties) (Jahn 1993). In the German case, the extent to which state EM has taken on board concerns as diverse as nuclear power and waste recycling is partly a reflection of the political pressure of Die Grünen under this electoral system.

The electoral system is only one way in which outside non-state groups influence state EM. The latter may also resort to the media, to social protest, or to public education campaigns aiming at changing environmental attitudes. Increasingly, such efforts are acquiring a global dimension through the development of "global civic politics" – that is, politics between environmental managers that do not necessarily implicate the state (Wapner 1995).

It is still the case today that states are influenced at the global level primarily by the interests and actions of other states, as witnessed by such political events as the Rio Earth Summit in 1992 (Grubb et al. 1993). Just as state EM is conditioned by factors internal to the country itself, so too such EM is subject to the growing impact of other states' policies and practices in an increasingly globalized world (Hurrell 1994).

In this regard, a key development in recent years has been the need for inter-state cooperation over a whole range of environmental problems that transcend national boundaries (Lipschutz & Conca 1993). Such cooperation has not only entailed the elaboration of international laws, conventions, treaties and agencies to deal with these "global" environmental problems (see Ch. 7), but it has also required significant changes in how states go about managing the environment within their own national boundaries. In order to conform with international agreements, states around the world have needed to modify policies and practices to bring them into conformity with such agreements. There has certainly been much rhetoric surrounding state compliance with international agreements (Caldwell 1990, Brenton 1994). In the 1990s, this process has been particularly associated with the worldwide push to reorganize state EM in line with the notion of sustainable development (Silva 1994). The latter concept is highly problematic for, as noted in Chapter 4, the discourse associated with this concept juxtaposes environmental conservation and economic development within the confines of the global capitalist system, without acknowledging the potential problems of this approach. This has not stopped states around the world from adopting this concept as a means to reform existing state EM practices, and thereby respond at least in part to the increasing concerns of the public (Bowlby & Lowe 1992). In this manner, the influence of the interstate system and that of non-state groups within a country may relate in such ways to represent an important influence on state EM.

Yet, not all states are equally influenced in their EM policies and practices by the inter-state system. Just as power relations are reflected within countries through the ability to shift uncertainty in EM from more powerful to less powerful environmental managers in society, so too power at the global level is reflected in the ability of stronger states to shift uncertainty in EM to weaker states. A notable case in point is the international movement of toxic waste from countries such as the USA and Germany to poor and politically weak recipient countries, as in sub-Saharan Africa. Such "waste colonialism" increases the hazards associated with EM in receiving countries while reducing environmental uncertainty in sending countries (Porter & Brown 1991).

More generally, the location of highly toxic manufacturing plants in many ELDCs is associated in part with the need for TNCs, who often operate such plants, to find host countries in which state pollution regulations are less stringent than in their home countries. Although this "pollution haven" thesis (Leonard 1988) must not be over-exaggerated, it is nonetheless a good example of the ways in which uncertainty in EM in a given country is a reflection in part of unequal power relations at the global level. Indeed, the so-called "Bhopal syndrome" – referring to the infamous release of toxic chemicals at a Union Carbide plant in Bhopal, India, in 1984, which resulted in deaths or injury of thousands of people – is only the most vivid manifestation of such a globalized "hierarchy" of environmental uncertainty (Weir 1988, Morehouse 1994).

The preceding discussion of the political aims and interests of the state has emphasized the significant but internally complex role of state EM and the various ways in which the state as an environmental manager is constrained or enabled by factors internal and external to the state itself. The professed objective of all state environmental managers is to enhance predictability in EM for society at large (Walker 1989). As the discussion of the EM implications of the internal organization of the state has highlighted, that goal may be undermined by the very way in which the state has been organized. In other words, the ultimate objectives of state EM and the institutional means by which the state seeks to pursue those objectives may be fundamentally at odds. Yet, if the state itself may, ironically, be an important source of uncertainty in EM, the existence of a diverse range of other non-state environmental managers may further complicate the quest for predictability in EM. Uncertainty may be inherent to the multi-layered character of contemporary EM.

Non-state environmental managers

The preceding discussion has highlighted the complexity of state EM. The existence of a diverse range of non-state environmental managers reiterates that EM is a multi-layered process in which environmental managers interact in the pursuit of their political interests and objectives. Yet, not all environmental managers share the same EM interests and objectives in their quest for predictability (see Table 2.2). This is reflected in their differing relationships to each other and to the environment.

At the periphery of the global capitalist system, hunter–gatherers have a set of objectives linked to their concern to maintain a distinctive and largely autonomous approach to EM. Theirs is a politics of marginality and an attempt to defend resource access and control. Such an attempt is also linked to preservation of a culture and way of life. Tactics of hunter–gatherers have typically been to retreat to ever more inaccessible areas before the onslaught of "development". Yet, such "avoidance behaviour" (Adas 1981) is increasingly difficult to sustain, as global capitalism now exploits even the remotest of regions. Even in Brazil's Amazonia – the largest remaining expanse of tropical forest – indigenous peoples are finding it increasingly difficult to maintain their EM practices in the face of encroachment from loggers, ranchers, miners and farmers (Treece 1990).

As avoidance behaviour increasingly loses its utility as a political strategy, hunter–gatherers must confront directly other environmental managers who are undermining their EM strategies (Hecht & Cockburn 1989). An increasingly important way in which they seek to do so is through organizations set up at the national or international level, often in conjunction with environmental NGOs (Broad 1993, Princen & Finger 1994, McCormick 1995). The case of the Penan of Sarawak illustrates the point. These environmental managers have been resisting infringement of their indigenous rights through various means, including road blocks designed to halt logging operations and through the publication of their ancestral domain claims in various national and international forums. In the latter endeavour, the Penan have received the critical support of such major environmental NGOs as Sahabat Alam Malaysia and the World Rainforest Movement (World Rainforest Movement & Sahabat Alam Malaysia 1989, Colchester 1993).

Farmers, who are more numerous and potentially more powerful than hunter–gatherers, are largely integrated into the global capitalist system. Their actions can influence EM at the national and even the transnational level (LeHeron 1988). Yet, as with the "state", it is important to understand the differentiated nature of farmers' political aims and objectives with regard to EM. Although it is possible to suggest that most, if not all, farmers around the world seek autonomy in their EM practices, in reality they constitute a highly differentiated type of environmental manager according to socio-economic status and environmental attitudes. This has implications for their ability to control decision-making on their land (Wilson 1996). At a general level, poor and often indebted farmers working small plots of land face greater constraints in managing the land according to their own aims than do wealthy landowners, who have both the financial means, and often political connections, to ensure substantial autonomy in EM decision-making (McDowell & Sparks 1989, Potter 1990, McEachern 1992, Wilson 1997).

Farmers do not always pursue their EM strategies through individual action, but join together in the guise of farmers' organizations to pursue collectively a common set of political aims and objectives. For example, the Welsh Farmers Union represents many farmers in Wales through lobbying and information activities. Such organizations can be powerful in the articulation of collective farmers' interests *vis-à-vis* conflicts with other environmental managers in society (e.g. environmental NGOs). This is highlighted, for example, through conflict between farmers' unions in the UK and some conservation organizations such as the Royal Society for the Protection of Birds over the management of remnant semi-natural habitats on farms (Lowe et al. 1986, O'Connor & Shrubb 1988).

The preceding discussion highlights the important point that farmers, as with many other non-state environmental managers, pursue their political aims and objectives, often, but not exclusively, in relation to the state. The objective is to promote predictability in EM practices for farmers through the promotion of autonomy in decision-making. Paradoxically, an important means of doing so is through the maximization of benefits obtained through the state, whether those benefits be financial resources or information (Robinson 1991, Whitby & Lowe 1994). Such an initiative, if successful, may not only shift uncertainty in EM onto other environmental managers, but also onto farmers in less powerful coalitions, regions or countries. Although farmers in the EU benefit from the Common Agricultural Policy, those elsewhere, especially in ELDCs, are thereby deprived of access to potential markets and subsidies that influence the nature and extent of EM practices associated with agricultural production (Robinson 1991, Robinson & Ilbery 1993).

The TNC is an increasingly important non-state environmental manager. Whether it be in the realm of taking from (resource extraction) or adding to (pollution; waste) the environment, these corporations have become one of the more important, but least understood, of modern environmental managers (Pearson 1987, Krol 1995, Welford 1996). In general, the primary aim of TNCs is profit maximization. To that end, these environmental managers have political aims and objectives typically associated with the avoidance of stringent regulations that may inhibit profit maximization. Such "avoidance behaviour" encompasses the evasion of regulations that seek to minimize the adverse environmental implications of resource extraction and manufacturing activities by TNCs (Leonard 1988). In this regard, many but not all TNCs seek to use their relative international mobility as a means to further these political objectives. As noted before, the relocation of TNC manufacturing to ELDCs is an example of this strategy. In this manner, and as highlighted in Chapter 2, TNCs seek primarily to promote financial predictability in their EM practices. In doing so, they often enhance uncertainty for other environmental managers in the host country through environmental degradation. RTZ and other large TNC mining companies have long sought to extract minerals at the expense of local habitats (i.e. deforestation, river pollution) upon which local environmental managers may depend. In Papua New Guinea, for example, large-scale mining (e.g. Ok Tedi Mine) has resulted in widespread environmental degradation down stream and social disruption of indigenous EM practices (Hurst 1990, Banks 1993). Similarly, the concentration of manufacturing enterprises that generate toxic wastes (e.g. asbestos) in urban ELDCs has created large urban pollution problems that adversely affect urban residents, and pose an increasing challenge for those state environmental managers who ultimately must address urban EM issues in these areas (Hardoy et al. 1992, White 1994).

TNCs seek to promote predictability in their EM operations through the avoidance of onerous state regulations and they also look to states for political and economic support in the establishment and smooth running of those operations. Just as farmers may seek financial and other support from the state, so too TNCs solicit the active assistance of states in their EM practices. Prior to establishment of operations, TNCs may bargain for the provision of state-funded infrastructure (i.e. roads, hydroelectricity, telecommunications) or other subsidies designed to render more attractive the corporation's activities within the host country. A case in point is the provision of a road network in the late 1960s and 1970s in Sumatra and Kalimantan (Indonesia) to facilitate logging by Weyerhäuser, Mitsui and other timber companies (Hurst 1990). Once established in a country, TNCs often maintain close links to host states with the aim of winning further support in the form of tax concessions and policy changes (Pearson 1987). Although relations between TNCs and states are not always harmonious, there has often been a convergence of interests around the maximum exploitation of people and environments in aid of the twin goals of national development and profit maximization.

In recent decades, international financial institutions have also become important environmental managers in multi-layered EM. Institutions such as the World Bank and the International Monetary Fund, and also regional institutions such as the Asian Development Bank and the Inter-American Development Bank, have played an increasingly influential role in shaping the EM policies and practices of other environmental managers (World Bank 1992,

Rich 1994). The relationship between international financial institutions and states, and especially EMDC states, has been especially close and it forms a powerful coalition in international EM decision-making. For example, the World Bank has promoted the EM interests of its EMDC sponsors (especially the USA) through policies and practices designed to influence state EM, notably in ELDCs (George & Sabelli 1994, Rich 1994). A case in point is World Bank support for the transmigration policies of the Brazilian and Indonesian states, with major implications for a range of state and non-state environmental managers (e.g. farmers; Hecht & Cockburn 1989, Rigg 1991).

This powerful "alliance" of state, IFI, and TNC interests has not gone unchallenged, and has often resulted in intense conflict (see below). More generally, this alliance has prompted the emergence of an increasingly assertive "counter-alliance" comprising environmental NGOs and other "grassroots" environmental managers such as farmers, fishers and hunter–gatherers (Rush 1991, Bullard 1993, Princen & Finger 1994). Unlike other environmental managers, environmental NGOs are not motivated by direct livelihood considerations, but couch their promotion of sustainable EM in language reflecting the idealistic and ecocentric political aims and interests that often underlie their activism (Bluehdorn 1995, Wapner 1995).

At a practical level, and reflecting their lack of direct involvement in EM practices, environmental NGOs seek to influence the EM policies and practices of other environmental managers in order to effect a change in those policies and practices consonant with environmental NGOs' political aims and objectives. Although environmental NGOs provide support and advice to farmers and hunter–gatherers, they also lobby states, international financial institutions and TNCs with the aim of modifying the EM activities of these environmental managers. If the impact of environmental NGOs on EM in general is indirect, it is nonetheless powerful especially in a world of globalized environmental problems that cannot be solved by states alone (Porter & Brown 1991, Laferriere 1994). In a sense, the development of environmental NGOs reflects a direct challenge to the state's claims to being the ultimate environmental manager; indeed, it may be seen as a condemnation of the state as an environmental manager responsible for the "wise stewardship" of the environment (Lowe & Goyder 1983). Therefore, the emergence of environmental NGOs has added to the complexity of multi-layered EM in which the political aims and interests of the environmental managers involved are more often than not in conflict.

Conflict in environmental management

Environmental managers do not operate in isolation from each other. As a result, EM is partly about how the political aims and interests of different environmental managers clash, thereby resulting in conflict. In some cases, though, these differences may form the basis for cooperative endeavours. Therefore, at the heart of understanding EM as a multi-layered process is the recognition that unequal power relations between environmental managers are reflected in both conflict and cooperation. The last section in this chapter addresses cooperation in EM, whereas the discussion in this section highlights the conflictual basis of multi-layered EM.

As noted, EM is conventionally seen as a technique to be used by state environmental managers to resolve potential or actual conflict over the environment. In contrast, the inclusive approach to EM adopted here emphasizes that conflict is a direct outcome of the often contradictory EM policies and practices of environmental managers. The latter approach sees conflict as an almost inevitable by-product of the existence of multi-layered EM. Conflict in multi-layered EM can be understood broadly in relation to both environmental characteristics and the characteristics of environmental managers themselves (Fig. 5.1).

As the figure highlights, there are various reasons why conflict arises in EM.

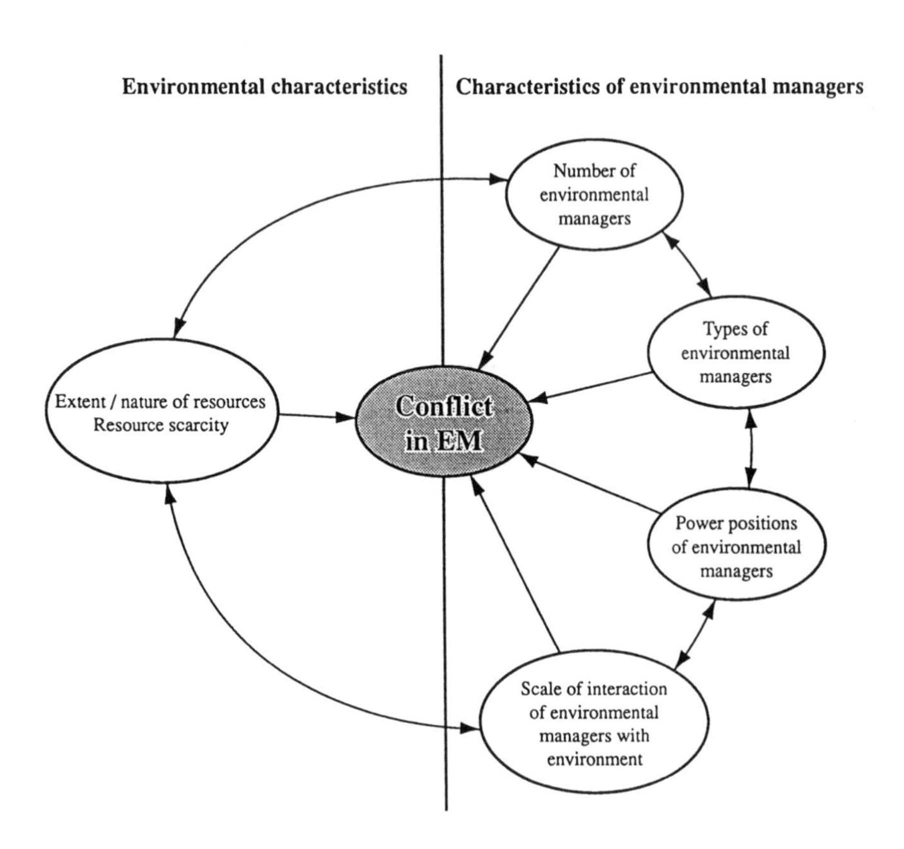

Figure 5.1 Conflict in multi-layered environmental management.

Perhaps most evidently, conflict may be associated with increasing pressure of population on resources or, put differently, with a growing number of individual environmental managers in a given area. For example, the migration of landless peasants in ever-increasing numbers from low-lying areas to upland forests in the Philippines has resulted in growing environmental degradation and conflict between migrant farmers and small populations of local shifting cultivators (Kummer 1992). However, such environmental degradation and conflict can occur "under rising PPR [pressure of population on resources], under declining PPR, and without PPR" (Blaikie & Brookfield 1987: 34).

Conflict is also associated with the competition for environmental resources by environmental managers pursuing various EM practices in the same area. In a given area, farmers or hunter–gatherers may pursue EM practices in accordance with their specific livelihood strategies. However, whereas the former seek to convert forest to field, the latter strive to maintain forests as the basis of their survival. As Harvey (1993: 29) argues, resistance by hunter–gatherers to intrusions from other environmental managers is often based upon a "recognition that an ecological transformation imposed from outside will destroy indigenous modes of production". Yet, it is often the impact of more powerful outside environmental managers that is associated with the most intense conflict in EM, especially in ELDCs (Turner 1989, Rush 1991).

In their drive to promote maximum production of commercially valuable resources, states have used their legal–political powers to devise policies of enclosure, whereby local environmental managers are excluded from a designated area as a preliminary step towards the full commercial exploitation of that area (*The Ecologist* 1993). For example, the initiation of mining or logging operations are typically associated with legal changes that privilege state EM over pre-existing indigenous EM practices (Broad 1993). In some cases, this process is initiated and implemented by the state, and more particularly specific departments within the state (Bryant 1997). The state frequently acts in conjunction with TNCs and international financial institutions in this process of resource exploitation. To take the Indonesian example again, Weyerhäuser and Mitsui were involved in extensive logging in Indonesia's Outer Islands at the behest of the Indonesian state, especially during the 1970s (Hurst 1990). For this reason, TNCs are often the focus of much opposition and condemnation by grassroots environmental managers (e.g. farmers) and environmental NGOs. At its most extreme, the intrusion of EM practices of states and TNCs into a given location may lead to the outright elimination of pre-existing local EM practices (Agarwal & Narain 1989, Shiva 1991). Grassroots environmental managers may migrate to other areas seeking new EM opportunities, but in an increasingly crowded and environmentally degraded world this is often not a viable strategy (Parnwell & King 1995, Sarre & Blunden 1995). As a result, these people often end up migrating to urban areas, and in the process typically lose their status as environmental managers.

To appreciate the nature and outcome of conflict between environmental managers, it is essential to understand how money, coercive capabilities and knowledge condition power relations between environmental managers. The financial resources of individuals and groups may play a pivotal role in determining the ability of an environmental manager to prevail in a given EM situation. For example, environmental managers can promote their goals through the cooptation of opposition: bribes to local politicians or government officials, offers of employment to rival environmental managers, or promises to "redevelop" the local environment to compensate for, or rectify, any environmental degradation that may occur. In Thailand, a national logging ban in 1989 notwithstanding, illegal logging by Thai companies has occurred with the tacit support of forestry officials (PER 1992).

The ability to deploy financial resources to promote EM goals is also often associated with the power to coerce recalcitrant local environmental managers. In this regard, the cooperation of political and economic elites needs to be emphasized (see below). TNCs and other powerful businesses possess considerable financial resources, but the state typically retains control over the means of coercion – the police and the army. Such means are often deployed in conjunction with business interests to promote large-scale EM practices such as logging or mining. In its most extreme form, this combination of business and state interests is reflected in a highly militarized and repressive approach to EM. For example, in such countries as Indonesia, Zaire and Brazil, large-scale EM projects that have resulted in widespread deforestation have been implemented only with the support of the armed forces (Hecht & Cockburn 1989, Witte 1992, Dauvergne 1993/4).

Control over environmental knowledge is another factor that conditions the power of environmental managers, and thereby influences the nature of conflict in EM. The state often has a considerable informational advantage over non-state environmental managers in that it commissions scientific research on both the biophysical and social aspects to EM. Such information and allied technologies (e.g. remote sensing, geographical information systems [GIS]), constitute a potentially powerful means to map, measure and manipulate people and the environment (Miller 1985, Breyman 1993, Pretty 1995).

To the extent that such information and technology is combined with financial and coercive means, the ability of powerful state and business interests to promote EM practices in the face of local popular opposition is enhanced. However, such power should not be overemphasized. First, state and business interests are not always synonymous. Indeed, conflict may even occur between these two environmental managers. This is epitomized, for example, through the conflict in the early 1990s over royalty rates and reforestation programmes between the New Zealand government and Japanese firms clearing native forest in New Zealand for wood chips (Memon & Wilson 1993, Wilson 1994b).

Secondly, and as discussed in the preceding section, the state itself is a highly complex environmental manager. Insofar as bureaucratic conflict may ensue from different state EM objectives, the ability of the state to use its coercive and informational advantage is thereby reduced. Conflict often arises between departments in charge of environmental conservation and those traditionally empowered to exploit the environment. For example, the New Zealand Department of Conservation is in frequent conflict with the state-owned Forestry Corporation over forest management issues (Roche 1990, Britton et al. 1992, Memon & Wilson 1993).

Thirdly, and despite the fact that many non-state environmental managers (i.e. farmers or hunter–gatherers) often do not have the financial, coercive, technological and informational capabilities of states or TNCs, they nonetheless may have important advantages that can be utilized in conflicts over the environment. To begin with, such environmental managers often have a very detailed understanding of the local environment; indeed, so-called "indigenous knowledge" often compares favourably in terms of local ecological understanding with state-commissioned "scientific" knowledge (Pretty 1995). Such knowledge can be used to powerful effect in conflict. For example, farmers and shifting cultivators can use it to resist the exactions of more powerful environmental managers through what Scott (1985) has termed "everyday forms of resistance". Such resistance is typically covert, anonymous, and performed by individual environmental managers. It includes such actions as "illegal" timber extraction, hunting or fishing. In India and Indonesia, for example, timber "theft" from state reserved forest is endemic and constitutes the essence of forest politics in these countries (Guha 1989, Peluso 1992). Such resistance is not designed as a direct challenge to state or other powerful environmental managers. Over many months and years it can nonetheless serve to undermine the EM policies and practices of those environmental managers. As Scott (1985: 36) notes, "just as millions of anthozoan polyps create, willy nilly, a coral reef, so do thousands upon thousands of individual acts of insubordination and evasion create a political or economic barrier reef of their own . . . it is only rarely that the perpetrators of these petty acts seek to call attention to themselves. Their safety lies in their anonymity".

Local resistance is gaining increasing power today as outside environmental managers, such as environmental NGOs, intervene to support local farmers and hunter–gatherers. Such outside intervention is based on the existence of democratic or quasi-democratic conditions in a given country; environmental NGOs, for example, have little opportunity to assist grassroots environmental managers in authoritarian countries such as Burma, Cuba or Libya. Even in democratic contexts, the role of outside environmental managers such as environmental NGOs is a politically sensitive issue; indeed, such participation may prove counterproductive. For example, insofar as state and business leaders are able to paint alliances between environmental NGOs and local environmental managers as examples of external interference in the internal affairs of a country, the ability of local environmental managers to resist state and business interests may be damaged. This may be the situation in Malaysia, where Prime Minister Mahathir has repeatedly attacked Malaysian resistance alliances for their international contacts (Eccleston 1996).

The power alliances between environmental NGOs and local groups may partially offset the ability of state and business interests to achieve their EM goals. In the Philippines, for example, 250000 farmers are organized as the Federation of Free Farmers to challenge state and business alliances (Pretty 1995). Further, it has been estimated that as many as one in ten Filipinos is involved with an environmental NGO or other grassroots movements (Broad 1993). Such alliances may be further strengthened through affiliations with international environmental NGOs. Those such as Greenpeace, Friends of the Earth and the Environmental Defence Fund provide financial, technical and informational support to local environmental managers throughout the world (Bramble & Porter 1992, Hurrell 1992). Perhaps more importantly, these environmental NGOs also draw media attention to otherwise neglected local environmental conflicts. Such media attention forces ostensibly more powerful environmental managers to justify their EM practices on the grounds of sustainable development and social justice. For example, the incident over the Brent Spar oil rig in the North Sea in the summer of 1995 pitted Greenpeace against Shell Oil over the proposed sinking of this rig in the deep ocean. As a result of large-scale media coverage, and a spiralling consumer boycott of Shell products, this TNC was forced to abandon its plans on environmental grounds – although it turned out that the grounds for this action were scientifically dubious.

Beyond the characteristics of environmental managers, conflict in multilayered EM is also related to environmental characteristics (see Fig. 5.1). In this regard, resource scarcity is often at the heart of conflict in EM. As Chapter 3 noted, the human impact on the environment has grown dramatically in recent centuries as a result of population increase and increased per capita consumption linked to the spread of capitalism. As a result, competition for environmental resources is becoming ever more fierce. Conflict will vary depending not only on the number and type of environmental managers involved but also on the extent and nature of the resource in question. In general, the more scarce the resource, the more likely is conflict. Intensity of conflict is also linked to specific characteristics of the resource itself; that is, conflict over resources vital to human existence (e.g. water) may be more intense than conflicts over nonessential resources (e.g. rubber trees). Perceptions of scarcity and livelihood necessity will nonetheless vary from place to place. What is perceived as scarce and/or essential to livelihoods of environmental managers in one locality may not be so perceived elsewhere (Mitchell 1989, Rees 1990, Mather & Chapman 1995; see Ch. 6).

Further, conflict operates at various scales, thereby highlighting the complex spatial interaction of different types of environmental managers and interests. As Chapter 3 emphasized, EM is concerned increasingly with the globalization of environmental problems. However, few environmental managers operate at the global level. However, those who do formulate policy at national or global levels are having an increasing impact in myriad localities.

Although conflict often occurs at the local scale, it needs to be understood in a broader context in which environmental managers removed from the immediate situation nevertheless can play an important role. The case of the Penan in Sarawak usefully illustrates the scale-dependent nature of environmental conflict. At one level, this conflict pits Penan hunter–gatherers against logging companies acting in conjunction with the Sarawak government. It has also been "globalized", in that the Penan struggle has been taken up by a wide range of environmental NGOs in both EMDCs and ELDCs (Hong 1987, Colchester 1993). As part of this process, the Malaysian federal state itself has become involved insofar as an internal conflict has become internationalized. In this manner, the Penan struggle is a local-level conflict related to livelihood needs, but has also become a symbolically important issue in a much wider conflict among environmental managers at various levels.

This section has highlighted the role of conflict in multi-layered EM. It has explored the conflicting power relations between various environmental managers and issues of resource scarcity and the multi-scale nature of many conflicts. Such conflict may be also conditioned by the existence of tactical alliances between various environmental managers in the promotion of shared EM interests.

Cooperation in environmental management

The existence of alliances among environmental managers illustrates that cooperation can also be an important factor in understanding the politics of EM. Paradoxically, such cooperation may be an integral part of conflict over environmental resources, in as much as environmental managers join together to fight over who shall control EM in a given area (see above). Cooperation may also be understood as an attempt to promote a broader consensus among competing environmental managers, which is to serve as the basis for a resolution of all outstanding differences. The ways in which cooperation among environmental managers may be expressed through the development of policies is considered in some detail in Chapter 7. Here, we examine briefly the political nature and meaning of cooperation for environmental managers as they pursue predictability in multi-layered EM.

What may be termed "tactical alliances" between environmental managers are designed to empower certain groups over other groups and they represent a common form of cooperation. The goal of such cooperation is to enhance predictability for alliance members through the transfer of uncertainty to less powerful environmental managers outside the alliance. Tactical alliances, therefore, represent cooperative situations that are nonetheless activated by a belief that EM represents a zero sum game.

The most common tactical alliance has linked together states, international financial institutions and TNCs in the pursuit of environmental resource exploitation. For example, Shell Oil has undertaken oil exploration and extraction in Ogoniland (southern Nigeria) with the full backing of the military government in that country, and accordingly has been able to pursue its business activities despite widespread local and international opposition (Rowell 1995). In recent years, this powerful tactical alliance has been increasingly challenged by a counter-alliance comprising environmental NGOs and diverse grassroots environmental managers (e.g. farmers, hunter–gatherers). This counter-alliance has developed to resist the interference of control of states, TNCs and international financial institutions while attempting to assert the primacy of community control over the environment (Fisher 1993). A case in point is provided by the cooperation between environmental NGOs, such as Friends of the Earth and the Environmental Defence Fund, and indigenous community groups, such as the National Council of Rubber Tappers and the Indigenous Peoples Union, the latter two groups representing many thousands of grassroots environmental managers in Brazil. The goal of this tactical alliance is to halt the destructive EM practices of an alliance between the state, businesses and international financial institutions – notably the Brazilian state, TNCs, local businesses and the World Bank – long held responsible for the promotion of environmentally destructive EM practices in the Amazon (Hurrell 1992, Bramble & Porter 1992).

Although tactical alliances have been the norm in cooperative relations among environmental managers, it is by no means certain that they enhance predictability for all participants. Indeed, the growing intensity of environmental degradation is such today that it is becoming clear that such alliances may be intensifying uncertainty in EM.

In contrast to tactical alliances, what may be termed "comprehensive alliances" promote cooperation among a wider range of environmental managers as part of an attempt to resolve uncertainty in EM. Comprehensive alliances are based on the idea that EM is not a zero sum game, and that environmental managers can benefit from cooperation through mutual accommodation and consultation (Young 1989). In other words, it is in the ultimate self-interest of all environmental managers to make selective concessions so as to enhance predictability for all concerned. However, there are political difficulties in developing such alliances associated with building goodwill and trust among a large number of participants.

The attempt to develop a comprehensive alliance in the 1970s to tackle the problem of extensive pollution in the Mediterranean Sea is a case in point. This initiative brought together a diverse range of environmental managers to attempt to develop a common understanding of the problem. The result was the Mediterranean Action Plan, developed at the Barcelona Convention in 1976, which committed participants to the reduction of pollution in the Mediterranean Sea. The innovative aspect to this alliance was that it involved the interaction of scientific advisers, state officials and local community leaders in building an "epistemic community" as a precursor to policy implementation

(Haas 1990). However, as this example also highlighted, there are difficulties in developing comprehensive alliances, especially in relation to the heterogeneity of EM interests involved. For instance, problems emerged as a result of political differences among participants. The Mediterranean example illustrates both the potential for cooperative behaviour and the obstacles to any effort to establish a comprehensive alliance.

Tactical and comprehensive alliances are not necessarily fixed. Indeed, these alliances are often in a state of flux in keeping with the shifting interests and priorities of participants. As Chapter 2 noted, environmental managers may emphasize different dimensions of predictability over others, even when faced by a common uncertainty (e.g. marine pollution in the Mediterranean). Under these circumstances, the politics of alliance-building is an extremely complex business. Environmental managers may join together in the pursuit of shared objectives in a given context, but they may later find themselves in opposition in other circumstances. For example, poor grassroots environmental managers may share with environmental NGOs a common interest in the elimination of environmentally destructive practices perpetrated by powerful environmental managers. Yet, grassroots environmental managers may later fall out with environmental NGOs as to the appropriate use of local environments. Although the latter may emphasize total environmental protection, the former may be more concerned with developing a sustainable livelihood from those environments, which nonetheless does not guarantee complete "protection" of the environment (Peluso 1993, Bailey 1995, Bryant & Bailey 1997). A further example relates to the dynamics of the state–TNC alliance over logging in Indonesia. In post-1967 Indonesia, the Suharto government initially encouraged direct foreign investment in the forest sector, with the result that American and Japanese companies, such as Weyerhäuser and Mitsui, began large-scale logging operations, notably in Kalimantan (Borneo). But beginning in the late 1970s, differences over whether to process the timber in the country (i.e. value adding before export) or to continue exporting unprocessed logs, as well as disagreements over the allocation of profits from logging, led to differences between the Indonesian state and the TNCs. In the end, the latter pulled out of Indonesia and have been replaced by indigenous firms closely linked to the government (Hurst 1990, Dauvergne 1993/4).

The preceding discussion of tactical and comprehensive alliances has illustrated briefly the complex interactions of environmental managers as these actors have sought to promote predictability through cooperation. One of the interesting features of this process is that state environmental managers often have needed to acknowledge the EM capabilities and activities of non-state environmental managers, especially those who already have considerable political or economic power. Cooperative efforts to enhance predictability for alliance members have therefore often necessitated a recognition by officials that the state does not have a monopoly on EM.

There are, nonetheless, many practical and political problems in devising alliances that can reconcile divergent interests. Indeed, conflict may exist not only between different types of environmental managers but also within each type of environmental manager (e.g. bureaucratic rivalry). This emphasizes how great the obstacles may be to cooperation in EM. This point will be returned to in Chapter 7 in the context of a discussion of the ambiguities surrounding policy coalitions.

Conclusion

This chapter has examined how politics may condition EM. It has suggested that, in a context of scarce environmental resources, power relations have an important influence on how multi-layered EM operates. Environmental managers have greater or lesser access to financial and other resources that can be used in EM conflicts. Further, environmental managers may cooperate with other environmental managers in the pursuit of shared interests. Such cooperation has typically been limited, with tactical alliances designed to assert partisan interests in a context of environmental conflict. Relatively few attempts have been made to develop comprehensive alliances encompassing the interests and participation of all interested parties as part of the promotion of predictability in EM.

This chapter has argued that political processes have scarcely enhanced predictability in EM. Indeed, these processes have exacerbated the uncertainty facing a range of actors, especially politically weak grassroots environmental managers. Political processes are both an opportunity and a constraint. Weak environmental managers have attempted to organize themselves, and often with the support of environmental NGOs, to assert their interests. Further, a growing recognition on the part of powerful environmental managers that tactical alliances do not necessarily enhance predictability for participants – and worse, may even lead to intensifying social and environmental uncertainty – presents an opportunity for a new politics of consensus based on comprehensive alliances.

If collective, rather than individual, predictability in EM is the ultimate goal, the need to understand how environmental conservation and the economic interests of environmental managers can be reconciled becomes a paramount concern. In other words, a new politics of consensus will need to grapple with the ways in which economic activity expressed through the market mechanism can be recast so that predictability in EM is promoted. However, the market in a global capitalist system often has been at the root of many of today's environmental problems (see Ch. 3). Paradoxically, and as the next chapter shows, the role of the market in EM is as essential as it is problematic.

CHAPTER 6

Environmental management and the market

A key factor in understanding the prospects for increased predictability in EM relates to the economic activities of different types of environmental managers as expressed through the market. To the extent that the market governs the livelihood prospects of most (if not all) environmental managers today, understanding its role in relation to multi-layered EM is imperative.

Environmental managers relate to the market in different ways, reflecting differing patterns of integration into the market and roles within it. Such differing patterns of integration and roles have important implications for the livelihoods of all environmental managers. This is especially the case within the current global capitalist system. Chapter 3 noted the historical development of that system and associated increased uncertainty in EM, but here attention focuses on how that system today produces "winners" and "losers" among environmental managers, and the attendant implications for the quest for predictability in EM.

Market characteristics and environmental management

The market is a means by which environmental managers exchange goods (and services) in the pursuit of their livelihood interests. The market has been associated with the policies and practices of environmental managers ever since those managers sought to exchange the environmental resources that they extracted or produced for other resources or services (Polanyi 1957, Marx 1973). Markets have operated under a succession of economic systems (e.g. feudal, socialist, capitalist), and have been influenced inevitably by the principles and practices of those systems (see Ch. 3). Five general issues need to be considered in seeking to understand the relationship between the market and multi-layered EM.

First, and perhaps most evidently, the market attaches a commonly recognized economic value to environmental resources. This value is dependent on the importance of the resource in question to environmental managers. That importance reflects various factors, notably supplies of a given resource, how essential that resource is to the livelihood needs of environmental managers, and issues relating to the cultural construction of resources and resource needs (see Ch. 4). The point here is not that all environmental managers agree with how the market values a given resource – far from it. Indeed, as noted below, dissatisfaction over market prices is a key feature of the current global capitalist market. Rather, it is to simply emphasize that the market has developed as a way in which to facilitate interaction between environmental managers in the pursuit of livelihood interests, through the creation of a uniform pricing system that regulates access to environmental resources (Rees 1990, Mather & Chapman 1995).

Secondly, the market provides a means by which environmental resources can be exchanged. It attaches a value to these resources through the medium of exchange – that is, the value of a given resource is ascertained through a continuing process of price adjustments based, in theory, on supply and demand. From an EM perspective, what is important here is the fact that the market facilitates the exchange of resources over large distances, thereby enabling the large-scale production or extraction of environmental resources in order to meet demand over a wide area (Eckersley 1996b). Under capitalism, such exchange through the market has assumed global proportions. For example, the integration of Bicol (a peninsula in south Luzon in the Philippines) into the global market placed a premium on the production of abaca used in the manufacturing of twine and rope for export, notably to Europe and North America beginning in the mid-nineteenth century. In the process, this led to the replacement of local market-orientated EM practices by farmers with those based on large-scale monoculture for the global market (Rush 1991).

Thirdly, and as a direct response to market valuation and exchange, the market has a direct bearing on the operation of multi-layered EM. Not only do environmental managers exchange resources in the market, they also often modify their EM policies and practices in order to try and maximize economic benefits gained through the market. However, just as different environmental managers possess different amounts of political power (see Ch. 5), so they also face different constraints and opportunities as they interact with each other through the market. Although the global capitalist market may be seen as the "great leveller", in that it requires all environmental managers to accept the exigencies of market fluctuations, more powerful environmental managers are nonetheless able to benefit disproportionately from market activity (see below).

Fourthly, the market serves as an important means by which resource scarcity is regulated. Scarcity illustrates how there can be a fundamental tension at times between the market and the quest for predictability by environmental managers. In general, scarcity can be linked to the physical availability of environmental resources. As Chapter 3 highlighted, the combination of rapid population increases and intensifying per capita use of, and impact on, the environment has placed growing pressure on selected resources (e.g. forests, clean water). Scarcity can also be a reflection of political and economic factors that may generate "imposed" scarcities. These may reflect a variety of motivations such as power, profit or prestige. The example of the Organization of Petroleum Exporting Countries (OPEC) is a prime instance of politically manufactured resource scarcity (Odell 1983). This cartel became influential in the early 1970s, as it restricted oil production in a move designed simultaneously to raise prices and to attack the USA economically for its support of Israel in the Arab/Israeli conflict. In the process "scarcity became a political issue" (Rees 1990: 31). Scarcity may also be culturally constructed, insofar as the availability of selected resources may be conditioned by prevailing environmental worldviews and attitudes of environmental managers (see Ch. 4). For example, the use of rhinoceros horn or crushed tiger bones for aphrodisiacs and "pharmaceutical" purposes in some Asian countries has intensified the demand for big game to the point where this "resource" has been largely depleted (Barbier et al. 1990, Peluso 1993).

Fifthly, externalities highlight what can be termed "market failure"; that is, that the market does not capture the full environmental implications of human–environment interactions. Indeed, as Rees (1990: 261) argues "externalities are the uncompensated side-effects of any economic or social activity . . . The word uncompensated is important as it serves to exclude all the external costs and benefits that arise in the course of normal market transactions". For example, long-term pollution is a process that the market left to its own devices has traditionally been unable to rectify (Burrows 1979, Eckersley 1996a). Such pollution is associated with the elaboration of industrial capitalism since the nineteenth century. Indeed, as Chapter 3 highlighted, such is the extent of pollution today that it has assumed global proportions (e.g. greenhouse warming, ozone depletion). Yet it would appear that the only way in which the market can account for such externalities is through intervention by state environmental managers. Thus, "free-market environmentalism" would seem to be an oxymoron (Eckersley 1993).

The discussion so far has illustrated briefly, and at a general level, the relationship between the market and EM. However, the market bears down on different environmental managers in different ways. It is essential to relate this general discussion to an analysis of how the market differentially affects the various types of environmental managers examined in this book. Similarly, it is essential to appreciate how the market may condition the policies and practices of those environmental managers. The next section examines the role and position of environmental managers in the context of the market, whereas the subsequent section evaluates more specifically how the global capitalist market generates "winners" and "losers" among those environmental managers.

Market integration and multi-layered environmental management

The relationship of different environmental managers to the market is complex, reflecting evolving market structures, as well as environmental changes that are linked to market operations. Table 6.1 sets out schematically the possible connections between key types of environmental managers and the market. This table does not purport to cover all aspects of the subject, but only those most pertinent to the themes of this book.

A starting point is to appreciate the general role of different environmental managers in the market. TNCs, farmers, the state and hunter–gatherers (to a limited extent), are all producers of marketable commodities. They take resources from the environment, such as timber, minerals, fish or cereals for sale in local, national, or global markets. The state nonetheless also serves as a regulator of the market, insofar as it seeks to address "market failures" relating to environmental degradation or social deprivation; it also intervenes in some instances (e.g. price-fixing for agricultural commodities) to ensure national self-sufficiency in food (Schramm & Warford 1989, Rees 1990, Eckersley 1996a).

The state is not alone in seeking to regulate the market. Both international financial institutions and environmental NGOs seek to regulate market activity in different ways. International financial institutions use their control over large financial resources (and their influence with private financial lenders) to condition production of diverse environmental resources for the market, in the process affecting the market supply of those resources (World Bank 1992, Rich 1994). Environmental NGOs, on the other hand, seek to influence the demand for selected environmental resources (e.g. mahogany, fur) as part of a larger campaign to end the trade in potentially endangered flora and fauna, at the same time that they often promote the sale of "eco-friendly" products (Friends of the Earth 1992, 1994a, 1994b, Wapner 1995).

Yet, environmental managers are not equal in the market. It is critical to appreciate the unequal power of different environmental managers as they become integrated into the market. In this regard, it is useful to distinguish broadly between environmental managers as "setters", "shapers" or "receivers" of market prices (see Table 6.1). At one extreme are hunter–gatherers, poor farmers and fishers, whose position within the market is one of price receivers. These groups have little or no control over the market price of the resources that they try to sell on the market. In contrast, well-off farmers or fishers may, through collective action (e.g. farmer lobbies), be able to shape demand and prices for selected resources (Goodman & Redclift 1991).

The case of states and TNCs provides an example of the other extreme, whereby environmental managers act as price setters in the market. However, both of these groups can also be price receivers insofar as the market is either too large, or competition is too great, for any one environmental manager to determine prices (e.g. TNCs and pulp & paper prices; see Marchak 1995). The record of market relations is replete with many examples in which states and

Table 6.1 Market integration and multi-layered EM.

	Role in the market	Potential power in the market	Degree of market integration/resiliency	Response to scarcity and externalities	Associated predominant EM practice
TNCs	• Producers	• Price setters • Price shapers • Price receivers	• Strong integration • Variable resiliency (depending on size, location, etc.)	• Contribute to scarcity and externalities • Seek to avoid regulation of activities contributing to scarcity and externalities	• Maximum exploitation • Little regard for conservation
State	• Producer • Direct regulator	• Price setter • Price shaper • Price receiver	• Strong integration • Usually resilient	• Contributes to scarcity and externalities • Seeks to regulate (directly/indirectly) scarcity and externalities	• Maximum exploitation • Some (and growing) regard for conservation
International financial institutions	• Indirect and direct regulators	• Indirect price shapers	• Strong integration • Usually resilient	• Indirectly contribute to scarcity and externalities • Indirectly seek to regulate scarcity and externalities	• Promotion of maximum exploitation • Growing attention to conservation issues
Hunter–gatherers	• Producers	• Price receivers	• Weak integration • High resiliency	• Adversely affected by scarcity and externalities	• Increasing pressure to alter often conservation-oriented practices
Farmers	• Producers	• Price receivers • Price shapers (e.g. large farmer lobbies)	• Moderate to strong integration • Low to moderate resiliency	• Contribute to scarcity and externalities • Adversely affected by scarcity and externalities	• Often forced to maximize exploitation
Environmental NGOs	• Indirect regulators	• Indirect price shapers	• Weak integration • High resiliency	• Indirectly seek to regulate scarcity and externalities	• Promotion of maximum conservation consistent with social justice

TNCs either shape or even set market prices for environmental resources. The case of OPEC was noted above, in which selected states, acting in conjunction with a small handful of TNCs, were able for a time in the 1970s and early 1980s to dictate the world price of crude oil (Odell 1983, Rees 1990).

The status of international financial institutions and environmental NGOs in terms of potential power within the market is different from that of other environmental managers, insofar as their role here revolves around attempting to shape prices indirectly. International financial institutions use control over financial resources to encourage other environmental managers to produce selected environmental resources, thereby potentially conditioning the market price of those resources. Loans by the World Bank to promote cash crop production in many ELDCs (linked to structural adjustment programmes) may constitute a powerful intervention in the market through boosted supplies of targeted resources (Rich 1994). Similarly, environmental NGOs seek to use their influence with consumers and other environmental managers to eliminate the trade in selected environmental resources. To the extent that their campaigns are successful in reducing demand for these resources, they play a potentially powerful role as indirect price shapers. A case in point is Friends of the Earth's (1994b) "mahogany is murder" campaign, which has disrupted the market in this timber through appeals to potential consumers and distributors.

In addition to the question of market role, the question of the potential impact of the market itself on environmental managers and their EM practices needs to be considered. In this regard, it is important to appreciate the degree of integration of environmental managers with the market and their resilience in the face of its vicissitudes (see Table 6.1). States, international financial institutions and TNCs are characterized by strong integration into the market insofar as their activities are closely associated with the market at the local, national and global scale. Some farmers and fishers share this strong market integration, but many others are somewhat less integrated into the market. For example, large New Zealand sheep and cattle farms produce agricultural commodities entirely aimed at the market (Wilson 1994). Many shifting cultivators, on the other hand, produce mainly for subsistence, with minimal production for the market (Pretty 1995). The latter are, therefore, weakly integrated into the market. Similarly, hunter–gatherers pursue their EM practices at the margins of the market economy, often opting wherever possible to remain independent from the market (although with growing difficulty). Finally, the market position of environmental NGOs is somewhat ambiguous, although the fact that they do not produce tangible commodities or services means that they are only weakly integrated into the market.

An additional factor that needs to be considered is the relationship between different types of environmental managers, market operations and associated environmental changes (see Table 6.1). The role of state environmental managers in the direct or indirect promotion of market activities that result in either resource scarcity or the production of externalities has been considerable. Yet, the state is attempting increasingly to regulate those activities to mitigate problems associated with growing resource scarcity and/or externalities (Weale 1992). A more indirect role is taken by international financial institutions, both in contributing to, and attempting to regulate, resource scarcity or externality problems. For example, loans provided to encourage environmental managers to plant trees or clean up polluting industry illustrate this role (Sargent & Bass 1992).

TNCs stand out here insofar as these environmental managers have made a massive contribution to both resource scarcity and externality problems. The impact of individual TNCs certainly varies, depending on size, location or corporate philosophy, as well as on the characteristics of specific industries (Pearson 1987, Welford 1996). It is evident that the increasingly globalized and footloose nature of TNCs is a powerful contributing force in environmental problems associated with taking from and adding to the environment (Korten 1995). Further, and despite business rhetoric, these corporations may seek actively to avoid regulations pertaining to their activities that contribute to resource scarcity or the production of externalities. Environmental NGOs, in contrast, are implicated in environmental problems linked to resource scarcity and externalities insofar as they indirectly seek to alter the practices of TNCs and other environmental managers in order to resolve these problems. Not surprisingly, therefore, environmental NGOs and TNCs often (but not always) find themselves in direct conflict over EM problems related to market activities (Princen & Finger 1994, Wapner 1995).

Of all the different types of environmental managers, it is the grassroots environmental managers who have typically been worst affected by resource scarcity and externality problems. Some of them certainly have actively contributed to these problems through unsustainable EM (e.g. depletion of fish stocks in many parts of the world), but the overriding impression is that fishers, farmers and hunter–gatherers are often severely disadvantaged because of the market-based activities of other environmental managers. For example, resource scarcity is typically associated with the depletion of resources essential to grassroots environmental managers, whereas externalities arising from the unsustainable EM of other environmental managers prejudice the ability of these grassroots environmental managers to pursue their own practices (e.g. water pollution affecting fish stocks).

Based on the preceding discussion of the market and multi-layered EM, it is possible to understand how market-derived motivations may animate the policies and practices of environmental managers. Although market forces are not the only explanation behind specific EM practices (see Chs 4 and 5), they often play a central role in conditioning those practices. In this regard, the market can be either an enabling or constraining force for environmental managers (see Table 6.1). On the one hand, states, TNCs and international financial institutions have typically gained wealth and power from their connections to the market. Of course, there are no guarantees in this regard, even for states, as the plight of some small and heavily indebted West African states illustrates. For example, most states have sought, directly or indirectly, to maximize exploitation of the environment; to reiterate a theme developed in Chapter 5, most states have sought to "develop" rather than to conserve the environment (Walker 1989). States are beginning to show a growing regard for the latter objective as they belatedly attempt to promote sustainable EM. International financial institutions have followed a similar pattern to that of states in that they have encouraged other environmental managers to maximize resource exploitation through market-based activities, with little regard to possible environmental degradation. As with states, some international financial institutions have begun to alter their EM policies to promote environmental conservation initiatives (Rich 1994).

As quintessential market participants, TNCs have fully exploited the environment for the production of commodities for the market. Typically, this has resulted in unsustainable EM, as resource extraction is often pushed to the limit to meet expected market demand (Korten 1995). These practices are reflected, for example, in the depletion of most of the world fish stocks by corporations (often abetted by states) bent on extracting as much fish as rapidly as possible, even in the knowledge that these stocks are increasingly limited (Fairlie 1995). The quest for profit by TNCs operating in a global market system also militates against them agreeing voluntarily to conservation measures (Korten 1995).

On the other hand, farmers, fishers and hunter–gatherers often have been forced either to adopt EM practices based on the maximization of resource extraction or have seen the territory within which they pursue their chosen EM practices whittled away by more powerful environmental managers. Hunter–gatherers, although often trying to persist with traditional EM, have been forced into ever-more remote areas because of the encroachment of other environmental managers into their territories in order to produce for the market (Hong 1987, Denslow & Padoch 1988). In contrast, those poor farmers and fishers already well integrated into the market have found that, as a result of market operations, their EM practices have been shaped increasingly by the necessity to maximize production for the market (Broad 1994). A case in point are poor and marginal New Zealand farmers who, during the 1950s, became integrated increasingly into global wool export markets and, as a consequence, had little choice but to expand their pasture areas at the expense of surviving remnants of native forest on their farms, in many cases resulting in severe soil erosion (Wilson 1993).

As the type of environmental manager least integrated into the market, the environmental NGO is not as caught up in the cycle of maximum exploitation and market operations as are others (see Table 2.2). Instead, environmental NGOs promote maximum conservation of the environment, contrary to the capitalist approach, as part of a broader effort to tackle resource scarcity and externality problems. Increasingly, environmental NGOs are linking this conservationist approach to calls for intragenerational social equity and justice, and in the process this has led many of these environmental managers to advocate fundamental market reforms consonant with such change (Princen & Finger 1994).

The preceding discussion set out the relationship of different types of environmental managers to the market in terms of their role, potential power and degree of integration with the market. The connections between these environmental managers, market operations, and prevalent EM practices was also generally evaluated. How different environmental managers interact with each other in the context of the global capitalist market, and how such interaction shapes their EM policies and practices, have yet to be considered.

Capitalism and environmental management: winners and losers

In Chapter 3, the relationship between the development of capitalism and intensifying use of the environment was discussed. Here, consideration is given to the issue of how capitalism as a system of economic organization tends differentially to benefit or penalize different types of environmental managers. The capitalist market now affects almost all environmental managers, but it does so in such a way as to create winners and losers.

At the heart of capitalism is the establishment of market relations according to the principles of profit-maximization. This profit-driven market has had immense implications for the interaction of environmental managers operating within multi-layered EM. The goal in this inherently unstable system is for environmental managers not only to earn a basic livelihood but also to accumulate as much capital as possible, with the aim of continually reinvesting that capital to maximize economic gains (O'Connor 1988, Johnston 1989). Under such a system and given increasingly scarce environmental resources, the operation of the market almost inevitably involves conflict between different environmental managers in the pursuit of profit. Indeed, at an aggregate level there is "a superexploitation both of the wealth of the earth and of human societies" (O'Connor 1994: 5).

More specifically, the capitalist market particularly rewards those environmental managers who can minimize production costs. This process has several dimensions. First, it encourages environmental managers to obtain environmental resources as cheaply as possible. This consideration has been a major driving force, for example, behind the gradual transfer of forest plantations for pulp and paper from EMDCs to selected ELDCs such as Thailand, Indonesia, Brazil and Chile (Marchak 1995, Lohmann 1996, Claro & Wilson 1996).

Secondly, the capitalist market encourages environmental managers to expand their operations in the first instance, and to employ any labour thereby hired as cheaply as possible in those expanded operations. Once again, the objective is to minimize costs and maximize profits. The relocation of many

TNCs from EMDCs to ELDCs since the 1960s has reflected in part a quest to locate production sites in settings characterized by low wage costs and "docile" labour forces (Pearson 1987, Korten 1995).

Finally, the capitalist market encourages environmental managers to avoid, whenever possible, regulations pertaining to those externalities created as by-products of the production process. To take the example of TNC relocation noted above, this process has occurred, in part, because host countries, most often in ELDCs, are seen as "pollution havens", especially by heavily polluting or toxic industries such as pesticide- or asbestos-manufacturing (Leonard 1988).

A key characteristic of the capitalist market is a process whereby more powerful environmental managers seek to exploit both their less powerful counterparts and diverse environmental resources essential to the production process. Yet, as O'Connor (1994: 5) observes, this process "does not entail mere appropriation of a surplus; rather, it is a destructive process whose result is the dereliction of human societies and ecosystems alike". A key aspect to this "destructive process" has been the restructuring of economic activity as part of a globalization of the capitalist system. Not only has market activity become increasingly synonymous with capitalism, it has also assumed global proportions. The general implications of this process have been noted in Chapter 3, but what is important here is to emphasize how this process has reinforced the tendency of capitalism to create winners and losers, but now on a global scale. The following discussion uses the examples of TNCs and small-scale farmers to illustrate this point.

TNCS are clear winners in the global capitalist system, whose development as an environmental manager has been an inextricable part of the emergence of that system. TNCs are able to expand their EM practices around the world, based on cost calculations of available supplies and production costs. These environmental managers are able to use the globalized market as a basis for maximizing the "efficient" use of resources and people, in that they are able to locate their production activities in the most propitious locations and yet make profits by selling their goods around the world (Korten 1995). The ability of these environmental managers to maximize advantage from the "global" reach of the market certainly varies from industry to industry (Pearson 1987). For example, mining TNCs are relatively constrained in their global EM policies and practices, depending on the availability of minerals, although the largest companies (e.g. RTZ) try to circumvent this potential problem through a policy of diversified mineral exploitation (Rees 1990, Moody 1996). Yet, for most TNCs the relative mobility afforded by the global capitalist system is a means to minimize production costs and associated state regulations so as to maximize profits (Korten 1995; see also "pollution havens", p. 93). A major EM implication of this process is that those who increasingly control EM decision-making are typically far removed from the potentially severe local environmental effects of any given EM decision. For example, decisions made by large chipmilling companies in Japan have caused the clearance of some of the last remnants of lowland native forest in remote parts of New Zealand, with associated biodiversity loss (Wilson 1994b, Marchak 1995). Thus, corporate decision-makers often do not see for themselves the environmental degradation ensuing from their EM policies.

In contrast, small-scale farmers (especially in ELDCs) have often been the principal losers in a global capitalist market system. The extent and nature of the integration of these farmers into this system has varied considerably. A general trend has been associated with the replacement of diversified (and often environmentally sustainable) production strategies, combining subsistence and selective market activity with the substitution of extreme market dependency on one or two cash crops (Pretty 1995). Chapter 3 noted the effects of the globalizing capitalist system on local-level environmental managers during the colonial era, but what needs to be emphasized here is the general point that the end result for most of these environmental managers has been financial hardship and increased uncertainty in EM.

Small-scale farmers certainly enjoy some financial benefits from the advent of a globalized market economy, notably in terms of an ability to buy hitherto unaffordable consumer goods with cash earned through their EM practices (Adas 1974). Yet, once integrated into that economy, small-scale farmers have often found themselves to be at the mercy of the market, with their livelihoods and EM practices subject to the vicissitudes of market activity (Adas 1974, Watts 1983a, Seabrook 1990). The tendency over time has been for these environmental managers to receive a declining financial return from their EM practices. In some cases, just to make ends meet, they have been forced into a desperate "ecocide" in which environmental degradation and poverty become a vicious circle (Blaikie 1985). For example, the fall in the price of a commodity such as groundnuts in the 1980s had implications in terms of soil-depleting EM practices resulting from a desperate need to maximize production to maintain incomes in selected West African countries (Franke & Chasin 1980, Watts 1983a). In the process, these environmental managers have typically lost the ability to control many aspects of their EM policies and practices, limiting to a certain extent their decision-making powers. Although the link, between poverty, market fluctuations and environmentally destructive EM practices is not necessarily straightforward (Broad 1994), poverty-stricken environmental managers have often been placed in a situation as a result of market forces in which sustainable EM practices are extremely difficult, if not impossible (Chambers 1987).

The preceding examples have illustrated how different environmental managers may gain or lose in the global capitalist market. An additional factor associated with that market is the implication it has for the state to fulfil its "stewardship" role. The shift to a global capitalist market has tended to undermine the ability of the state to manage the environment. Chapter 3 noted how the growth of that market was linked to the promotional efforts of states, and that the states have benefited from this globalization process (Walker 1989). As the "contradictions" of capitalism become increasingly apparent at all scales, the ability of states to protect the environment is diminished to the extent that global market activity is based on a relative absence of state regulations (O'Connor 1988). In many parts of the world today, the sheer pressure to maximize profit overpowers the efforts of even the best-intentioned of states in this respect. For example, efforts by the Costa Rican state to demarcate a significant proportion of the national territory as protected areas have been largely thwarted because of the pressures associated with operating in a capitalist system. National park boundaries have been encroached upon as a result of expanding logging and cattle-ranching activities (Utting 1993). Thus, although states may seek increasingly to protect the environment, they find that they are largely unable to do so because of the globalized capitalist market. In this regard, states too may be seen as "losers" under the current system.

The market does not operate in a neutral fashion. Rather, it rewards or penalizes different environmental managers, depending on the resources and power that they bring to the market. In this manner, it is a means by which uncertainty in EM can be transferred from more powerful to less powerful environmental managers. Yet, in an era of "market triumphalism" (Peet & Watts 1993), there would appear to be little feasible alternative to a market-based solution to the world's global environmental problems. Whether such a solution can ever lead to enhanced predictability in EM is a crucial question.

Enhanced predictability in environmental management through the market?

The discussion in this chapter, as well as in Chapter 3, has illustrated the growing importance of the capitalist market in the operation of multi-layered EM. The argument so far has been that this process has led to growing uncertainty in EM for most, but not all, environmental managers. Can predictability in EM ever be attained through the mechanism of a capitalist market? The following discussion reviews the conflicting views that exist on this question to attempt a general assessment of the relationship between the global capitalist market and predictability in multi-layered EM.

One view is that a global capitalist market left to its own devices will serve ultimately to adjust the policies and practices of environmental managers in such a way as to enhance predictability for these managers. In this context, the argument of "free market environmentalists" is that most, if not all, of the environmental problems facing humankind today are not attributable to the capitalist market per se, but rather to interference by other environmental managers – above all, the state – in what the state considers to be the smooth functioning of the market mechanism (Simon & Kahn 1984, Beckerman 1995).

For example, it is suggested that environmental externalities do not reflect a weakness on the part of corporate environmental managers operating in a global capitalist market, but "are attributed to the absence of markets and property rights in relation to the environment; if property rights over the environment are well defined, then the problem can be addressed through voluntary transactions among those causing environmental degradation and those suffering such degradation" (Eckersley 1996b: 15).

In this argument most environmental problems today may be attributed to "policy failure" rather than market failure (see also Ch. 7 on the "the tragedy of the commons"). The problem for environmental managers is not "too much market" but "too little market" presence in their EM practices. To take one of the arguments, problems of resource scarcity are seen to be best dealt with through a free global capitalist market unencumbered by regulations. As a resource becomes scarce, its monetary value will tend to increase, leading to the ever-more intensive quest for resource substitutes, thereby enabling contemporary EM practices to continue (Simon & Kahn 1984).

These arguments have been widely contested by those who suggest that the global capitalist market is intrinsically unable to enhance predictability in EM. It is suggested that there is no inherent control mechanism within capitalism that would encourage environmental managers producing for the market to peg production at a level compatible with the principles of sustained yield and carrying capacity, let alone with the precautionary principle. Indeed, the relative lack of control mechanisms often results in exploitation beyond sustained yield (e.g. forest destruction), and in the use of resources beyond carrying capacity (e.g. through overstocking). It also means that there is little incentive to follow the precautionary principle, as this would entail reduced profits (see Ch. 2). Further, and as Chapter 3 noted, the growth of the externality problem has been closely associated with the spread of capitalism around the world. Profit – which drives the capitalist system – is often based on the fact that capitalists do not have to pay for the diverse forms of pollution that are an inevitable by-product of their activities.

A second view has been that the best way in which to promote predictability in EM would be to reform the market mechanism in such a way as to put a price on hitherto unpriced "environmental goods". The problem here is that the present capitalist system treats many essential ecological functions (e.g. air, water) as free goods, leading to their seemingly inevitable degradation. In this view, the market may be a potentially useful means to promote environmental policies and practices by a wide range of environmental managers, consistent with enhanced predictability in EM. As set out by the likes of Pearce et al. (1989) and Barbier (1993), the task, therefore, is to suggest ways in which the market can incorporate hitherto unpriced environmental goods so as to encourage modified EM policies and practices conducive to sustainable EM (e.g. DuVair & Loomis 1993, Kuitunen & Tormala 1994).

This point has been illustrated with reference to suggestions that "national income accounts" – which summarize the financial implications of the market at the national level – need to be overhauled. These income accounts give an incomplete picture of the environmental implications of EM practices expressed through the market. Paradoxically, national income accounts often represent resource extraction, which may lead to environmental degradation, as a net increase in income. To take logging, for example, under the current system only the financial gains to be derived from commercial logging appear in these accounts. In contrast, reform-minded environmental economists suggest that a different system might offset such gains by noting the costs associated with possible environmental degradation (e.g. loss of biodiversity, watershed degradation) (El Serafy & Lutz 1989).

In this view the need to attach a monetary value to essential, but hitherto unpriced, ecological functions is suggested as the best way in which to correct "market failure" (Turner 1993, Bromley 1995). At the heart of these efforts to value the environment is the assumption that placing a monetary value on environmental goods is an essential prerequisite for rational decision-making in EM (e.g. Dwyer 1986). Such efforts assume that the market must necessarily be extended into new areas. Further, this view tends to suggest that, in order to integrate the environment fully into the market, the state must play a leading role – whether it be through direct intervention or via indirect means (i.e. regulation of business) (Sarkar & McKillop 1994). Although acknowledging the weaknesses of the global capitalist market system as presently structured, this view suggests that, suitably reformed, that system can serve as an effective means to promote predictability in EM.

A third view rejects the former two suggestions, arguing instead that predictability in EM and the global capitalist system will never be compatible. In effect, this view suggests that the capitalist system is based on environmental degradation and the social deprivation of weaker environmental managers (O'Connor 1988). At the heart of this view is the assertion of an essential connection between the global capitalist system, unequal power relations and environmental degradation (Johnston 1989). Under this system, the goal of environmental managers typically revolves around the quest to accumulate capital as rapidly as possible. In order to do so, they seek to minimize production and labour costs in a process that simultaneously involves maximizing the use of environmental resources at a minimal outlay. The specific form that this process may take varies with the type of environmental manager. For example, farmers may decide to use chemical inputs to boost yields, but in doing so may contribute to long-term environmental degradation, thereby leading to a decrease in predictability over the long term (Briggs & Courtney 1989). Similarly, and as mentioned above, TNCs seek out inexpensive environmental resources for their production activities, which often entails the adoption of environmentally degrading activities, once again prejudicial to long-term predictability in EM (Korten 1995, Moody 1996). The point here is not that the global capitalist market leads to identical EM practices in multi-layered EM. On the contrary, it establishes a basic disposition among environmental managers to favour environmental exploitation over conservation.

This third view suggests that there is a fundamental "contradiction" between the global capitalist market and sustainable EM (Redclift 1987). Accordingly, all efforts to reform that market are ultimately doomed. Rather, the objective should be to devise an alternative economic system which replaces the quest for capital accumulation with a recognition of the need to reconcile social equity and environmental conservation concerns (Pepper 1993).

The preceding discussion has highlighted that considerable differences exist over whether it is possible to promote predictability in multi-layered EM within the confines of the present capitalist market. However, as this book has made plain at various stages, this market is intimately associated with ubiquitous uncertainty in EM at the local, regional and global scale. Given the vested interests of powerful environmental managers who influence the nature and direction of the global capitalist market, the prospects for a reconciliation of that market with sustainable EM appear remote.

Conclusion

This chapter has examined the relationship between the market and multi-layered EM. It has suggested that the market is important in EM decision-making insofar as most environmental managers derive their livelihoods through interactions with the market, and respond to changing market demands and prices accordingly. Yet, the argument was made that the market is not a neutral process in which all environmental managers are equal participants, but rather a mechanism that tends to reinforce unequal power relations in EM. As Table 6.1 suggested, environmental managers are integrated into the global capitalist market in a highly differentiated manner in line with their power status, and the market power of environmental managers could be indicated in their relative ability to set or influence market prices.

Such unequal power relations are reflected in the global capitalist market. This market is based on capital accumulation, which tends to accentuate economic inequalities among environmental managers, and typically leads to environmental degradation. The historical record tends to suggest that capitalist development and environmental degradation are closely linked (see Ch. 3), but there are, nevertheless, divergent opinions as to whether a suitably reformed global capitalist market is compatible with increased predictability in EM for all environmental managers. The weight of the evidence presented so far in this book would tend to indicate that the prospect for such a reconciliation is unlikely.

The preceding two chapters have tended to emphasize that contemporary political and market structures are associated with increasing uncertainty in

EM, since these structures are seemingly based on unequal power relations and environmental degradation. It is important to consider to what extent these structural forces are reflected in the actual policies and practices of environmental managers, which is the subject of Chapter 7.

CHAPTER 7
Environmental management and policies

To understand how environmental managers go about actively and self-consciously manipulating the environment, it is important to appreciate the nature and dimensions of the policy process associated with EM. In order to analyze efforts to promote predictability in EM, not only do we need to appreciate politics and the market (see Chs 5 and 6), but also how environmental managers interact with the environment; that is, the policies that they devise and seek to implement in their EM practices.

Policies can be broadly understood as a framework for action developed by environmental managers in keeping with their quest for predictability. Traditionally, environmental policies have been equated with the activities of selected state agencies. In keeping with the inclusive understanding of EM adopted in this book, policies are seen here as being devised and implemented by all types of environmental managers. Following discussion of the different policy dimensions to multi-layered EM, the chapter will then examine the attempts by state and non-state environmental managers to devise environmental policy. In the final section, the development of policy "coalitions" is explored to highlight how the different policies of environmental managers may be combined in multi-layered EM.

Policy dimensions in multi-layered environmental management

The discussions of politics (Ch. 5) and the market (Ch. 6) have illustrated the key ways in which environmental managers interact in multi-layered EM in the quest for predictability. In contrast, an examination of the different dimensions to policy-making highlights how environmental managers attempt to specify strategies as a means to pursue diverse policy goals in relation to the environment itself. In this regard, there are four dimensions that need to be addressed (Table 7.1).

First, the ways in which environmental managers attempt to devise and implement policies is conditioned in part by how they go about understanding the environment they seek to manage. Environmental managers must develop some understanding of the environment if they are ever to develop environmental policies. Yet, knowledge of the environment is not simply a prerequisite for action. Rather, it also conditions the very ways in which such action occurs

Table 7.1 Policy dimensions.

	Policy-making			
	Scale	Timescale	Knowledge construction	Audience
Hunter–gatherers	Local	Long-term	Spiritually sanctioned appreciations	Family/local community
Farmers	Local	Medium-term	Oral tradition and positivist Western science	Family/local community
TNCs	Global	Short-term	Data accumulation based on positivist Western science (e.g. environmental impact assessment, GIS)	Shareholders/company owners and employees
International financial institutions	Global	Short-term	Data accumulation based on positivist Western science (e.g. environmental impact assessment, GIS)	EMDC states
Environmental NGOs	Global	Long-term	Data accumulation based on positivist Western science (e.g. environmental impact assessment, GIS)	Humankind/national citizens
State	National/global	Electoral terms	Data accumulation based on positivist Western science (e.g. environmental impact assessment, GIS)	Citizens

(Szerszynski et al. 1996). To a certain extent, what is possible in EM is a reflection of what is understood by environmental managers as the key traits and potential of the environment to be managed.

As Table 7.1 illustrates, environmental managers construct knowledge of the environment in different ways depending on diverse cultural, political, and market factors. The differing epistemological bases[1] of EM can vary considerably between environmental managers. For example, many hunter–gatherers and shifting cultivators have devised environmental policies in tune with spiritually sanctioned appreciations of the environment (Olofson 1995). In a world perceived to be largely controlled by powerful and potentially life-threatening spirits, EM for these environmental managers is partly a matter of subsistence and survival, but also partly a question of spiritual redemption (Hong 1987, Bryant 1994a). In contrast, state agencies develop environmental knowledge largely through the accumulation of data according to the principles of positivist Western science (Pretty 1995, Gumbricht 1996). From this perspective, the environment is known through efforts to describe and quantify its constituent parts (Miller 1985). In recent centuries, it is the latter epistemological approach that has been in the ascendancy because of the globalization of Western capitalist production and cultural values (see Chs 3 and 4).

1. The cultural and knowledge-based background within which environmental managers construct an understanding of their environment.

Secondly, policy-making by environmental managers is influenced by the intended audience, and, thirdly, by the scale of the EM issues themselves (see Table 7.1). The approach taken by farmers will be conditioned by the fact that these environmental managers are concerned with establishing policies that pertain to a specific plot of land, and in relation to a specific set of concerns linked typically to their families. In contrast, the state as an environmental manager seeks to devise environmental policies reflecting the nationwide extent of its formal responsibilities. For the state, and its specific agencies, the concern is not necessarily with any one plot of land or family, but with more general social and environmental considerations pertaining to the polity as a whole. Recently, the rise of transnational environmental NGOs, such as Greenpeace and Friends of the Earth, has led to a situation whereby these new environmental managers appeal to a global audience with calls for policy formulation that go beyond the preserve of the state (Princen & Finger 1994, Wapner 1995).

Fourthly, efforts by environmental managers to devise policies are also influenced by the different timescales on which environmental managers operate. Some environmental managers are more inclined than others to operate solely within short-term decision-making timeframes. In particular, political leaders, influenced by electoral considerations, typically need to think about policy-making in short three- to seven-year terms (Lowe & Goyder 1983, Lyon 1992). Insofar as they are driven by short-term considerations of profit maximization, TNCs can also be seen to operate predominantly in comparatively short timeframes (Hutchinson 1995). In a similar vein, international financial institutions tend to operate within short-term timeframes, reflecting their need to satisfy the financial and ideological interests of the USA and other leading EMDCs (Rich 1994). In contrast, Table 7.1 illustrates that many farmers seek to devise policies that, although not necessarily addressing future generations, nonetheless are based on medium-term timeframes encompassing several decades (Ward & Lowe 1994, Morris & Potter 1995). Finally, and given the opportunity, many hunter–gatherers have sought to establish environmental policies that place a premium on long-term intergenerational considerations. In this regard, these environmental managers are not so different from leading environmental NGOs, which also emphasize such considerations, although the latter act with regard to the wellbeing of the planet as a whole (Denslow & Padoch 1988, Princen & Finger 1994).

It is important to acknowledge that important differences often occur within each type of environmental manager with regard to policy formulation and implementation. For example, whereas many political leaders think in the short term, other leaders adopt medium- to long-term considerations. Indeed, the same person may adopt different timescales for different environmental policy issues. In the case of Norway, for instance, Prime Minister Gro Harlem Brundtland has been a firm backer of national and global long-term environmental initiatives (WCED 1987); yet, Brundtland has also been at the head of a government that stands accused of supporting practices leading to the extinction of whale species (Stoett 1993). More significantly, those state environmental managers who operate within a bureaucratic context, such as forest officials or fisheries officers, often tend to promote environmental policies based on medium-, if not long-term, thinking. These state environmental managers do so in part because the nature of their day-to-day work means making resource decisions with medium- to long-term implications (e.g. establishment of a forest plantation), but also because their employment situation (i.e. relative job security) may allow them to elaborate long-term strategies that also help secure their positions. With regard to the latter point, the promotion of medium- to long-term EM practices builds in a demand within the state to perpetuate funding and employment beyond the electoral timeframe that most commonly defines politicians' EM decision-making. Meanwhile, many farmers tend to base their environmental policies on medium-term thinking, but heavily indebted farmers are often forced into adopting policies that are based on short-term thinking out of sheer necessity (Blaikie 1985, Broad 1994).

Environmental managers may be influenced by differing epistemologies, timeframes, audiences and scales in the development of environmental policies. As Chapters 5 and 6 have emphasized, however, environmental managers do not operate in isolation from each other. Rather, they are caught up in often complex relationships based on political and market-related factors. Such factors, in turn, condition the ability of environmental managers to pursue the environmental policies of their choice. Since EM as a whole is a multi-layered process, it follows that the environmental policies that individual environmental managers adopt will partly reflect this situation in terms of multiple and overlapping environmental policy approaches.

Policy-making characteristics

As a result of differing interests and opportunities, environmental managers often adopt different policies as they seek to "actively and self-consciously manipulate the environment". In assessing the policies of state and non-state environmental managers, the goal-orientated nature of such policies is central. Later in this chapter, the ways in which environmental policies are devised and implemented in the context of multi-layered EM are evaluated. Here, the objective is to explore the policy-making characteristics of different types of environmental managers.

The state

The role of the state – even in a re-evaluated EM – remains central to environmental policy-making (e.g. Dorney 1987, Buckley 1991, Nath et al. 1993). Yet, the state is not the only policy-maker in EM. Hence, what is important to consider here is the distinctive nature of the policy-making role of the state as an environmental manager *vis-à-vis* other environmental managers.

In this regard, it is useful to distinguish between the state's direct and indirect policy-making roles (see Table 7.2 below). Just as with a farmer or fisher, the state through its various agencies is typically involved in devising policies linked to the direct and active manipulation of the environment. Forest officials, for example, directly manage and exploit state-owned forests. In contemporary Burma (Myanmar), for example, the Forest Department establishes the parameters of forest use and extraction, whereas the Myanmar Timber Enterprise is the state corporation responsible for teak and other commercial logging of the country's tropical rainforests (Bryant 1997). Similarly, park officials in many parts of the world implement policies covering all aspects of EM in those areas designated as national parks. In Kenya, for instance, the Kenyan Wildlife Service (a state agency) is responsible for implementing policies, designed to protect biodiversity, that, among other things, have resulted in the displacement of local environmental managers such as the Maasai tribe (Peluso 1993).

Clearly, the extent of policies based on direct state manipulation of the environment varies spatially and temporally. For example, in the former communist Soviet Union and East European countries, such manipulation was extensive and it encompassed everything from natural resource extraction to heavy industry (Waller & Millard 1992). In contrast, in the fervently capitalistic UK of the 1990s, policy-making associated with direct intervention plays an increasingly minor role in the context of progressive privatization of state-owned environmental assets such as water companies or electricity generation utilities (Jordan 1993). Yet, state EM policy-making through direct intervention differs from comparable non-state EM policy-making in that only the former occurs on behalf of the "common" good within the national territory.

However, it is the state's indirect EM policy-making role that in many cases is the more important. As many political theorists highlight, the distinctive feature of the state as an environmental manager is that it has a monopoly on the means of coercion in a given territory (Skocpol 1985, Mann 1986; see Table 7.2 below). From an EM perspective, this is important, because it means that the state – alone among environmental managers – is formally in a position to coerce non-state environmental managers in the pursuit of its own EM policy (again formally designed for the "common" good; Johnston 1989). As with other environmental managers, the state certainly seeks to influence the EM policy-making of others through non-coercive means. Voluntary land management agreements are a case in point. For example, agri-environmental schemes in the EU, notably the Environmentally Sensitive Areas scheme, involve state agencies in efforts to persuade farmers through diverse financial incentives to adopt more environmentally friendly EM practices (Baldock et al. 1990, Whitby 1994).

It is typically through the use of formal coercion (e.g. laws enforced by the police) that the state seeks to impose environmental policies on non-state environmental managers. The range of such indirect state environmental policy-making is often vast, reflecting the diversity of environmental issues and problems confronting contemporary society (see Ch. 3). For example, state policies designed to regulate logging on private and publicly owned land are enforced through legal acts that require non-state environmental managers to adjust their EM practices in line with rules defined by the state. In New Zealand, for instance, the recent introduction of the Resource Management Act 1991 curtails the freedom of farmers to clearfell or otherwise modify native forests on their land (Robertson 1993, Wilson 1994c). In other cases, state environmental policies are designed to manipulate the polluting practices of environmental managers responsible for industrial production. National policies in the USA associated with the federal Environmental Protection Agency are perhaps the most highly developed in this regard, and encompass a complex system of emission controls in which firms may trade off stricter controls on some plants for more lenient controls on other company plants (Hays 1987, Rees 1990).

In this manner, state environmental policy-making is designed to manipulate the EM practices of others associated with both taking from and adding to the environment. Further, in countries increasingly characterized by what Beck (1992) has termed "risk societies", the state's environmental policy-making role – supported by state-of-the-art scientific expertise – plays a crucial role in shaping both the EM policies of non-state environmental managers, as well as the ensuing environmental risks to all the citizens under its jurisdiction. In many areas, such as the regulation of nuclear energy use and allowable pesticide limits, states seek to control the EM practices of other environmental managers in order to attempt to minimize the health and environmental risks associated with potentially lethal substances (Blowers 1993).

In seeking to develop environmental policies, state environmental managers are aided by various policy-related tools. Most notable in this regard is environmental impact assessment which is a method designed to identify the likely environmental impact of planned projects (Wathern 1988). Although heavily criticized for its simplistic assumptions (e.g. Hollick 1981, Cocklin et al. 1992), this method is nonetheless a means of information gathering that is designed to help state environmental policy formulation and implementation. Other techniques, such as remote sensing or computer-based GIS, similarly provide a basis for state environmental managers to develop management plans as they seek to manipulate the environment indirectly (Cross et al. 1991).

The distinctiveness of the state's policy-making role is that it holds formal responsibility for the promotion of the "common" good within a national territory, and that it has a formal monopoly on the means of coercion within such a territory to fulfil that function (Table 7.2). This helps explain why many traditional accounts have tended to exalt the environmental policy-making role of the state (e.g. Dorney 1987, Buckley 1991). Yet, as the following illustrates, the policy-making role of non-state environmental managers can also be highly significant.

Non-state environmental managers

Although lacking the state's formal coercive powers and mandate, non-state environmental managers nonetheless develop environmental policies that provide the framework for their EM practices. Table 7.2 highlights the policy-making characteristics of different types of environmental managers. It shows that there are not only differences in policy-making characteristics between state and non-state environmental managers, but also among non-state environmental managers themselves.

Table 7.2 Policy-making characteristics of environmental managers.

	Policy influence	Formal coercive power	Examples of policy mechanism	Main rationale for policy	Intergenerational equity
Hunter–gatherers	Direct	No	Ancestral domain (mainly oral)	Subsistence	Yes
Farmers	Direct	No	Farm management plans (oral and written)	Subsistence and profit maximization	Yes
TNCs	Direct	No	Corporate environmental statements (written documents)	Profit maximization	No
International financial institutions	Indirect	No	Environmental guidelines (written documents)	Stability of the capitalist system	No
Environmental NGOs	Indirect	No	Policy statements (written documents, media)	Common good	Yes
State	Direct and indirect	Yes	Policy statements; environmental laws and regulations (written documents, media)	National common good and capital accumulation	Yes (officials) No (politicians)

For hunter–gatherers and shifting cultivators, "policy-making" comprises the development of often intricate "policies" that are designed to enhance predictability in EM. Such policy-making has traditionally not been articulated through written documents. Rather, policy has been expressed through oral traditions or myths. For example, oral traditions of the G/wi bushpeople of Namibia reinforced the notion that angering their "Supreme Being" by over-exploiting environmental resources would lead to environmental degradation (Simmons 1993). Similarly, folk tales of the Dogon people of Mali specify in considerable detail the procedures associated with, and limits to, the feliing and use of trees (Van Beek & Banga 1992). Recently, and often with the help of environmental NGOs, hunter–gatherers and shifting cultivators have sought to express their environmental policies in written and even cartographic form as part of an attempt to assert often wide-ranging EM claims against the state and other environmental managers (Peluso 1995). In the Philippines, for example, the Environmental Research Division of the Manila Observatory has been undertaking work on the island of Mindanao to help local indigenous Lumad people to stake claims to community forests through "community mapping" techniques (Braganza et al. 1994).

A central part of this latter endeavour is the assertion of policy-making based on the notion of ancestral domain (Colchester 1993). This notion, which directly contradicts state claims to control the national territory, asserts the right of the indigenous environmental manager in question to have autonomy in environmental decision-making in a given area. Typically, ancestral domain claims are linked to a comprehensive set of policies designed to facilitate the management of local environments in an integrated and holistic manner. Using modern mapping and other techniques, the goal here is also "to appropriate the state's techniques and manner of representation to bolster the legitimacy of "customary" claims to resources" (Peluso 1995: 384; see also Braganza et al. 1994). In contrast to the state, such policy-making reflects the direct interaction and dependence on the environment of hunter–gatherers and shifting cultivators – that is, these environmental managers devise policy in line with location-specific goals rather than the more general goals associated with, for example, the state.

In some cases, these goals are associated with carefully planned and timed migratory movements – the case of the Maasai in Kenya is illustrative. It was the traditional practice of these environmental managers to follow migratory paths carefully calibrated to the availability of fodder resources in different areas (Little 1987). However, nomadic pastoralists typically face even greater hurdles than do hunter–gatherers and shifting cultivators, who operate within relatively well defined and spatially concentrated territories. These relatively mobile people have policies that necessarily are based on multiple and often complex negotiations with the different environmental managers (including the state or states) that manage the environment in those areas encompassed by their pastoral uses (Bassett 1988).

For farmers, policy-making is conditioned by the fixed boundaries of the land that they manage and the typically more simplified ecosystems that they have created for their livelihood (i.e. fields). For many farmers, policy-making is predominantly associated with putting together a management package based on maximizing food and fibre production through the manipulation of their land, notably through the addition of chemical or organic inputs (Briggs & Courtney 1989, Pretty 1995). Such policy-making needs to consider external factors such as market prices and state policies. As with hunter–gatherers and shifting cultivators, farmers are primarily concerned with devising and implementing policy at the local level (see Table 7.1 above). Unlike these environmental managers, farmers typically are much more integrated into the global capitalist economy. They tend to be more concerned with basing policies on the goal of profit maximization – as is the case notably with TNCs and other businesses linked to EM (see below). It would be wrong, though, to equate farmers' policies with food and fibre production alone. For example, many farmers in the EU now devise policies that seek to incorporate conservation of semi-natural habitats on their farms. Farmers in southern Germany are a case in point. These environmental managers have developed their own policies, aimed at safeguarding locally important juniper moorlands (Wilson 1995). In these cases, policy-making is an exercise in which conservation and production goals are reconciled in the context of a comprehensive and integrated land-management plan devised by the farmers themselves.

Farmers' plans will vary considerably in complexity depending on local circumstances and the size of the holding. They will nonetheless typically include provisions for the coordinated use of land and inputs. For example, where conservation of certain habitats is the goal, policy will encompass such things as tree-planting, long-term fencing programmes and protection of wildlife (Wilson 1992a, Pretty 1995). On the production side, policy will be involved with the minute regulation of chemical or organic inputs, water use, field rotation (allowing land to lie fallow), or contour ploughing (to avoid erosion). Farmers' policy formulation, therefore, can take many forms. In many cases today, it will be represented through documents that specify long-term conservation and production goals, and linked to this the detailed practices that are required to meet such goals. However, many small-scale farmers undertake "policy formulation" in a less formal and usually unwritten style. In this case, policy formulation is associated with oral traditions (Chambers et al. 1993).

In contrast to farmers and hunter–gatherers, TNCs base their environmental "policies" on national, if not global, considerations (Hutchinson 1995, Brophy 1996). As Table 7.2 shows, they share with these environmental managers a direct association with the resources they use and upon which they depend. TNC policies encompass myriad concerns: for example, labour costs, resource availability, market proximity, and the extent of environmental regulations (Pearson 1987, Leonard 1988). The overriding concern for TNCs is profit maximization, and their environmental policies are ultimately based on this concern. As a result, and to a greater extent than any other environmental manager, TNCs are concerned with matching production levels to ensure maximum monetary returns. This could mean simply a "cut-and-run" policy, that is, a plan based on maximizing production levels in the shortest possible space of time. Yet, this need not always be the case. For example, TNC policies may be based on the limitation of production to drive up market prices and hence profits. These policies will reflect the complexities of not only market operations (see Ch. 6) but also the environmental policies of other environmental managers, especially the state (see Ch. 5).

Further, and to counteract their typically "bad" image as environmental managers, TNCs are resorting increasingly to policies based in part on conservationist criteria. For example, RTZ as one of the world's leading mining corporations, makes much of its efforts to restore the landscape subject to its mining operations (Rio Tinto Zinc 1994, Moody 1996). Paradoxically, the relative mobility of many TNCs at a global level means that, unlike other non-state environmental managers, they ultimately do not possess a long-term "stake" in environmental conservation in any given area. Indeed, these "rootless" and largely unregulated environmental managers may have largely superseded pre-existing EM decision-making structures in many parts of the world (Korten 1995).

Unlike the environmental policies of many small-scale farmers, hunter–gatherers and shifting cultivators, those of TNCs are largely encapsulated in written documents. Such documents range from corporate mission statements to shareholders, through to internal strategic plans, right down to location-specific operational guidelines. TNCs as diverse as Shell (oil), RTZ (mining), Weyerhäuser (logging) and Carskill (agriculture) now claim to base their policies on sustainable EM strategies (Eden 1994, Hutchinson 1995, Marchak 1995).

As with TNCs, international financial institutions develop their policies in line with global issues, but unlike TNCs and other environmental managers, these environmental managers do not directly interact with the environment (see Table 7.2). Yet, as with states (to which they are linked), they enjoy considerable policy-making influence with a wide range of other environmental managers. The mechanisms through which that influence is exercised relate to the withholding or transferral of large sums of money, or, more importantly, the approval or rejection of the environmental policies of states, which has an immediate bearing on the willingness of private banks to provide loans to governments. Institutions such as the World Bank or the International Monetary Fund have been widely criticized for an indirect policy-making role linked to massive and widespread environmental degradation (George & Sabelli 1994, Rich 1994). In response to such criticism, the World Bank created an environmental department in 1987 designed to ensure that World Bank policies were based upon sustainable EM (LePrestre 1989). In this manner, the World Bank, as with other international financial institutions, has acknowledged in effect that, although it does not directly manage the environment, it nonetheless has a highly significant, if indirect, environmental policy-making role (World Bank 1992).

International financial institutions articulate their environmental policies in a manner similar to the state or TNCs. The World Bank, for example, has clear environmental guidelines associated with all the loans that it now dispenses. In cases where recipient states refuse to adhere to those guidelines, the loans are withheld. A notable case in point is that of the Narmada Dam project in central India, in which early World Bank participation was subsequently withdrawn because of a local and international furore over the adverse social and EM implications of this project (Rich 1994). The World Bank also has specific sectoral policies. For example, it has recently finalized a forest policy that will serve into the next century as the basis for decisions concerning loans that have a bearing on sustainable forest management principles.

To a large extent, environmental NGOs also operate on the basis of an indirect policy-making role. The environmental NGO community is large and diverse, and in such a heterogeneous community there are environmental NGOs that manage the environment directly. For example, English Nature owns and manages small areas of land and water, and has policies that apply directly to these areas. Yet, in many cases the policy-making role of environmental NGOs is indirect (as with international financial institutions), insofar as policies aim to persuade and cajole other environmental managers to implement sustainable EM practices (see also Fig. 1.1). In this respect, environmental NGOs publish policy statements designed to persuade states, TNCs, community groups and even environmental users to modify their behaviour in line with sustainable EM principles. For example, Friends of the Earth produces a handbook for new members that serves not only as an introduction to the philosophy and aims of that organization but also to publicize more widely the long-term policies of this transnational environmental NGO (e.g. Porritt 1990). Further, environmental NGOs mount targeted campaigns through the media to increase the political pressure on influential environmental managers reluctant to heed the policy pronouncements of environmental NGOs. Recent cases in point include the Greenpeace-led boycott of Shell Oil products during the Brent Spar episode, or the Friends of the Earth "mahogany is murder" campaign, designed to shut down a trade that was reputedly contributing to environmental degradation and the genocide of indigenous peoples in Brazil's Amazonia (Friends of the Earth 1994b).

Although many environmental NGOs share with international financial institutions a largely indirect policy-making role in EM, these two types of environmental managers differ in terms of their ability to see their policies to fruition. Whereas international financial institutions use their control over financial resources (and indirectly the financial resources of private banks) to persuade other environmental managers (i.e. states) to modify their practices, environmental NGOs rely on their ability to summarize and publicize the ethical bases for, and the necessity of, sustainable EM. Environmental NGOs depend not on money but on the power of ideas in a world increasingly anxious about the globalization of environmental problems (Princen & Finger 1994). In the Brent Spar episode (summer 1995), the ability of Greenpeace to force one of the largest and most powerful TNCs in the world to modify its EM practices (despite the support of the British state for Shell Oil's original position) is a good example of the growing ability of environmental NGOs to influence EM practices of other environmental managers (Wapner 1995). Through high-profile campaigns and published documentary evidence, environmental NGOs articulate environmental policies which they believe are in the collective interest of humanity as a whole.

No environmental manager is in a position to act in isolation from other environmental managers. The development of environmental policies therefore needs to be situated in the context of multi-layered EM. To some extent, managers will acknowledge such interdependency in modifying their policies in anticipation of the policies of others. Clearly here, power relations are in operation. For example, the environmental policies of small-scale and politically weak farmers often need to be adapted to take account of environmental policies of more powerful environmental managers, such as large and politically well connected farmers or TNCs (Scott 1985, Little & Watts 1994).

Yet, it is typically the relationship between the state (with its monopoly on formal coercive powers) and non-state environmental managers that often has the greatest bearing on the formulation and implementation of environmental policies of individual managers. State policies often require that other environmental managers modify their practices in line with specific emission thresholds. A case in point is the imposition of air quality guidelines that necessitate the installation of complex and costly filters in factory chimneys. For example, the passage in the USA of the Clean Air Act in 1970 (and subsequent amendments) mandated the Environmental Protection Agency to establish national ambient air quality standards for specific pollutants. These standards ultimately required TNCs and other industrial environmental managers to adhere to national emission standards necessitating businesses to modify their existing EM practices (Hays 1987, Rees 1990, Miller 1994).

As Chapter 5 highlighted, there may be sound reasons for environmental managers to seek to harmonize their policies with one another from time to time. The following discussion of policy coalitions is designed to illustrate the ways in which environmental managers may seek to cooperate over policy issues.

Policy coalitions

Multi-layered EM means that there are many overlapping policies promoted at the same time by environmental managers. Yet, as Chapters 2 and 5 illustrated, the pursuit of predictability by one type of environmental manager may overlap with similar efforts by other environmental managers, thereby often resulting in cooperative endeavours. It is this potential but selective commonality of interest that provides the rationale for the possible development of "policy coalitions" between different types of environmental managers. The following discussion explores the ambiguities surrounding policy coalitions.

Policy coalitions reflect the concern of environmental managers to pursue predictability in multi-layered EM often characterized by unproductive conflict. Chapter 5 briefly explored the political aspects to cooperation in EM. Here, it needs to be emphasized that any policies that result from such cooperation may be themselves predicated on the differentiated quest for predictability by environmental managers (see also Ch. 2).

A policy coalition in the UK that addresses agricultural concerns illustrates this point. Over the past 50 years, a reasonably durable coalition has developed around questions of agrarian production and environmental conservation. For a long time, emphasis was given to maximizing commodity production in order to promote national self-sufficiency and export earnings. In recent decades, concern over mounting agricultural surpluses, as well as the increasing environmental degradation caused by intensive agriculture, has prompted a "greening" of agricultural policy (Potter 1990, Robinson 1991, Whitby 1994). These shifting policy concerns may be associated with the differing EM interests of those environmental managers involved in formulating policy. Although the primary concern of farmers (as represented by the National Farmers Union) has been the pursuit of financial predictability, the interest of the state (represented by the Ministry for Agriculture, Fisheries and Food, and key political figures) has emphasized concerns about political and sociocultural predictability (e.g. electoral success, inexpensive good-quality food for the public; Marsden et al. 1993, Whitby & Lowe 1994). In recent years, selected environmental NGOs (e.g. the Countryside Commission) have asserted the importance of environmental predictability in the deliberation of this policy coalition (Whitby 1996). Notwithstanding the fact that different environmental managers have emphasized varying dimensions of predictability, there has been enough common ground for the interested parties to come together and develop an overall policy package generally acceptable to all participants.

The development of policy coalitions has often taken place within the context of the nation state. The usual pattern has been for state agencies to join with other environmental managers to attempt to develop a common policy relating to a specific environmental issue. The role of the state has often been crucial in terms of attempting to build policy consensus (Lowe & Goyder 1983, Marsden et al. 1993). That role has also been significant in transnational efforts at policy formation.

A prominent example in this regard is that of the Common Agricultural Policy (CAP), which links together member states of the EU through a policy framework aimed at regulating agricultural production and countryside conservation. The CAP was formulated under Article 39 of the Treaty of Rome in 1957 by the six original member states of the European Community, but has since been expanded to encompass an extended EU comprising fifteen member states. To date, the CAP is the most comprehensive transnational policy effort in Europe (Robinson 1991, Robinson & Ilbery 1993, Whitby 1996), and has been shaped by the interests of a variety of different types of environmental managers. These include state officials linked to national agriculture ministries, farmers' unions generally representing more powerful farmers in member states, and selected national (e.g. Bund für Umwelt- und Naturschutz Deutschland in Germany) and international environmental NGOs (e.g. Birdlife International). Yet, what differentiates the CAP policy coalition from national policy-making frameworks is that it has also been shaped by supranational environmental managers linked to the Commission of the European Communities in

Brussels (in particular Directorate General VI). The latter aim at promoting diverse dimensions of predictability at the transnational, rather than the national, scale; for example, environmental predictability related to water pollution prevention between countries, or sociocultural predictability through support of agriculturally marginal holdings in less favoured areas (Potter 1990, Whitby 1996).

Compared to most other attempts at policy coalitions at the transnational scale, the CAP forms one of the few policies that binds its coalition members to the regulations set out in the policy framework. A case in point is the recently formulated "agri-environmental regulation" 2078/92/EEC which obliges all EU member states to establish agri-environmental schemes aimed at promoting enhanced environmental conservation in the countryside within their national boundaries (European Commission 1992). Although the response of individual members states to the regulation has varied considerably, member states were nonetheless forced to draw up and implement schemes (e.g. Environmentally Sensitive Areas) that compensate farmers for financial losses incurred through the adoption of more environmentally friendly EM practices (Wilson 1995, 1997, Whitby 1996).

Although national and transnational policy coalitions, as discussed above, may provide an effective basis for translating cooperation among environmental managers into policy practice, the creation of these coalitions does not necessarily mean that the concerns and interests of all environmental managers affected by the policies are considered. Policy coalitions do not necessarily overcome unequal power relations between environmental managers (Young 1989). Indeed, they may even reinforce them insofar as coalitions increase the ability of certain environmental managers acting together to pursue predictability at the expense of others excluded from the coalition.

To take once more the agricultural policy example in the UK, whereas that coalition often has been able to assert the interests of its members, it has nonetheless failed to represent the special interests of many small-scale farmers whose livelihoods are nonetheless potentially affected by decisions made by the coalition. Indeed, small farmers may be worse off as they may not always be in a position to benefit from the financial incentives associated with policies formulated by powerful policy coalition members (Lowe et al. 1986, Wilson 1996). Similarly, and with regard to the CAP, certain participants in this policy coalition have benefited more than others. With regard to some of the agri-environmental regulations that are part of the CAP, for example, Whitby (1996: 237) has argued that implementation "was mainly confined to Northern European Member States whereas the Mediterranean countries, for various reasons, were slow to take advantage of that opportunity". Although the CAP may have partly alleviated discrepancies between farmers' incomes within national boundaries – through Less Favoured Areas designation, which provides finance to keep farmers in agriculturally disadvantaged areas, for example – it may have nonetheless increased the rift between northern and southern EU member states (cf. Garrido & Moyano 1996). These two examples illustrate the general point that policy coalitions may be prejudicial to the interests of weaker environmental managers.

The preceding discussion has emphasized the potentially important role of national and transnational policy coalitions in the pursuit of predictability in EM. These coalitions are nonetheless ambiguous, in that they may favour the interests of some environmental managers over others. Ambiguities surrounding policy coalitions are nowhere more evident than in the global context, where selected environmental managers come together to attempt to resolve global environmental problems.

Global environmental problems have assumed growing prominence in recent years and efforts to develop policy coalitions in the global arena have been similarly emphasized (Young 1989). In the absence of a global state or Leviathan, collective action may be fated to fail in the context of a "tragedy of the commons"[1] in which conflict is often the norm among environmental managers (Hardin 1968). Unlike national contexts, the development of global policy coalitions is especially difficult because of the absence of a unified political territory ruled by a single actor.

To understand the distinctive problems that surround global policy coalitions, it is important to explore the implications of the "global policy vacuum" that has long characterized international EM (Young 1989). Hardin's (1968) "tragedy of the commons" metaphor is a useful way in which to appreciate the traditional paralysis in EM concerning global environmental problems such as global warming, ozone depletion and ocean pollution. Central to this concept is the idea that actors will seek to maximize their personal benefits from "common" environmental resources (taking from), and that conversely they will seek to minimize the costs associated with polluting activities (adding to). The "tragedy" results from the fact that, in the absence of an overriding authority, it is in every actor's interest selfishly to exploit the environment beyond critical environmental thresholds (Keohane & Ostrom 1994). The tragedy of the commons can be interpreted as the global manifestation of multi-layered EM. As Chapter 3 noted, contemporary global environmental problems reflect a long legacy of EM practices that have contributed to pervasive environmental degradation. Such degradation has assumed global proportions, as manifested by the accumulation of pollutants in the oceans and atmosphere (Cunningham & Saigo 1992, Pickering & Owen 1994).

This process needs to be related to the distinctive interests and activities of key types of environmental managers. For example, the rise of TNCs partly reflects the existence of a global policy vacuum. The environmental policies of TNCs are formulated usually to maximize the benefits to be gained from operating in this vacuum (Korten 1995; see Table 7.1). In the absence of worldwide

1. This is the tragedy that ensues because of lack of control of over-exploitation of common resources, such as the atmosphere, open oceans, biodiversity and Antarctica.

sets of emission standards, for example, TNCs may locate manufacturing operations in countries with low emission standards. Although, as noted in Chapter 5, the "pollution haven" thesis must not be exaggerated, most heavily polluting industries are now located in ELDCs, where national pollution regulations are typically weak (Leonard 1988, Hardoy et al. 1992). Under such circumstances, a "double tragedy" ensues – a tragedy often for poor local people living near manufacturing sites (e.g. Bhopal disaster in India), and a tragedy for the global environment as such virtually uncontrolled emissions may contribute to global environmental degradation.

In contrast, transnational environmental NGOs develop policies that try to overcome the global policy vacuum. Indeed, the meteoric rise of such environmental NGOs as Greenpeace and Friends of the Earth since the 1970s is based on popular anxiety associated with the perceived "tragedy of the commons" (McCormick 1995). A prominent policy concern for them relates to global issues such as ocean pollution and wildlife depletion (e.g. whaling). In many cases, environmental NGOs come into conflict with TNCs – conflict that reflects the differing policy rationales of these two types of environmental managers in the context of the tragedy of the global commons (Wapner 1995).

States play an ambiguous role in relation to the global policy vacuum. As this book has shown, states pursue national interests through their EM policies. Yet, in doing so, they often contribute to global environmental problems. For example, many ELDC states attract TNCs to their countries through lax environmental policies, thereby often contributing to increased environmental degradation. States also represent a leading obstacle to the promotion of global environmental policies because of their insistence on the pre-eminence of national sovereignty as a basis for most of their environmental initiatives (Mische 1989). States are, nonetheless, still seen today as pivotal environmental managers in efforts to devise a global policy response to the tragedy of the commons (Young 1989). This reflects a growing recognition among state leaders that the pursuit of national interests will inevitably entail the quest for at least a minimal set of global policy objectives – that is, that global policy coalitions bringing together states and other key environmental managers (e.g. TNCs) is a prerequisite for the pursuit of national interests by states (Vogler 1995).

It is the recognition by key environmental managers that intensifying global environmental problems require the development of global policy coalitions that explains why growing efforts are being made to overcome the serious hurdles to global cooperation (Young 1989, McCormick 1995, Rowlands 1995). Chapter 5 highlighted at a general level the politics associated with the interaction of different types of environmental managers operating at the international level. Here, it is useful to explore in more detail the ambiguous policy outcomes of such interaction as manifested through fledgling global policy coalitions.

States have played a prominent role in the development of global policy coalitions that are designed to address selected global environmental problems. The usual procedure has been the organization of international conferences as a mechanism by which to elaborate global policy coalitions, with the policy output of such meetings often taking the form of international treaties and agreements that aim to guide the activities of signatories.

Although there have been many international conferences on diverse environmental matters, it is with the United Nations Conference on the Human Environment [Stockholm Conference] in 1972 and the United Nations Conference on Environment and Development [Rio Conference] in 1992 that the state-led collective response to global environmental issues is perhaps best known. Despite the divisions and acrimony associated with these two major conferences, certain global policy decisions were derived. At the Stockholm Conference, for example, agreements were signed concerning marine pollution, Arctic seal culling, and the creation of a specialist UN agency (UN Environment Programme) dedicated to the promotion of global sustainable EM (McCormick 1995). Similarly, at the Rio Conference there were agreements on global biodiversity, climate change, forest management and Agenda 21, which were collectively designed to serve as the basis for nationally orientated sustainable management policies (Grubb et al. 1993, Middleton et al. 1993).

The efforts to develop a global policy coalition in order to address the problem of stratospheric ozone depletion perhaps best illustrates the difficulties that may arise when diverse environmental managers attempt to reconcile policy differences at the global level. A policy coalition dealing with this environmental problem began to develop in the late 1970s as a result of growing fears that the emissions into the atmosphere of chlorofluorocarbons (CFCs) and other substances were leading to the depletion of the Earth's ozone layer (Benedick 1991). This process is associated with an increase in the level of harmful ultraviolet radiation, which reaches the surface of the planet with potentially disastrous implications for life on Earth. Initial impetus for this policy coalition was derived from the work of scientists, which specified the rapidity with which ozone depletion is occurring. In the 1980s, states began to cooperate more closely with each other, with TNCs linked to the production of CFCs, and with environmental NGOs whose campaigns targeted this problem, culminating in the Convention for the Protection of the Ozone Layer (the so-called "Montreal Protocol") signed in 1987.

Although this policy agreement was accepted by 56 states in 1987, considerable opposition to the agreement developed in many ELDCs – led by China and India – over the question of compensation for foregone industrial production linked to ozone-depleting emissions (Miller 1995). The concession in 1991 by the USA, and other leading EMDCs, that ELDCs should be compensated under the terms of the Montreal Protocol, facilitated this policy's wider acceptance among concerned states (McCormick 1995). The Montreal Protocol denotes a clear recognition by states (and implicated TNCs and environmental NGOs) that environmental predictability needed to be the top priority in relation to the ozone question. As a result of this agreement, production of ozone-depleting

CFCs and other substances has declined. As such, it appears to be a relatively successful example of the formation of a global policy coalition.

The prospects for success of state-led global policy coalitions must not be exaggerated. The willingness of states to commit themselves to global policies is linked to the extent to which the proposed policy is issue-specific and not likely to generate political opposition within signatory countries (Hurrell 1994). Success tends to occur when relatively few signatories are involved and when the policy does not necessitate significant changes in existing EM practices (e.g. Montreal Protocol). Yet, many global environmental problems today require the involvement of a wide range of environmental managers, and the introduction of major political and economic changes to the status quo. In the process, the development of a policy coalition becomes exceedingly difficult. For example, collective agreement concerning the issue of global warming has yet to be achieved, mainly because of the daunting political obstacles associated with curtailing emissions across the globe (Clayton 1995).

Further, the success of state-led policy coalitions may be limited by the partiality of the state as an environmental manager itself. Most states are reluctant to surrender their EM powers to a collective global entity, and only rarely have they reflected the interests of the different environmental managers they claim to represent. Indeed, and as Johnston (1992: 226) argues "anarchy thus rules: most states are prepared to enter international agreements when they are in their interests . . . but not to establish bodies with powers over their nationals and their interests". The ambiguous role of states in relation to the development of policy coalitions is thereby emphasized.

Global policy coalitions have also been attempted by non-state environmental managers acting separately from the state. These efforts may reflect disenchantment with state initiatives designed to tackle global environmental problems. For example, environmental NGOs – often acting in conjunction with representatives of grassroots environmental managers – have sought to develop global policy coalitions designed to alter human–environment interaction in various ways. As with state-led policy coalitions, environmental NGO-led initiatives have often used the mechanism of environmental conferences. Beginning in the 1970s with the Stockholm Conference, and gaining momentum in the 1980s and 1990s, "alternative" conferences have brought together environmental NGOs and grassroots representatives. Although often keen to criticize the perceived shortcomings of state-led global policy coalitions, alternative coalitions have usually also sought to formulate global policies that incorporate the interests of grassroots environmental managers (Chatterjee & Finger 1994).

The campaign that decries destructive logging practices and promotes local "extractive reserves" in the Brazilian Amazon is a case in point. Environmental NGOs, such as Friends of the Earth and the Environmental Defence Fund, have long condemned the invasion of land in Amazonia belonging to grassroots environmental managers – and no more vociferously than at the Rio Conference. Grassroots environmental managers have organized themselves to fight for local control over environmental resources, including the "sustainable" production of a diverse range of non-timber forest products (Hecht & Cockburn 1989). These organizations include the National Council of Rubber Tappers and the Indigenous Peoples Union. What makes the policy coalition comprising selective environmental NGOs and grassroots environmental managers a global concern is the way in which a worldwide campaign has been mounted by these various actors, designed to pursue the goal of local EM for indigenous people in Amazonia. As noted above, Friends of the Earth has mounted a "mahogany is murder" campaign in the UK designed to persuade consumers not to buy mahogany, thereby potentially eliminating through "consumer power" those EM practices that threaten grassroots livelihoods in the Amazon (Friends of the Earth 1994b). In this manner, an alternative global policy coalition has developed in order to promote major political, economic and sociocultural changes that, it is argued, will assist grassroots environmental managers. The goal is to promote political and environmental predictability through the assertion of local EM powers and opportunities.

In contrast, leading TNCs have joined together to form a global policy coalition concerned with global environmental problems, but which is essentially supportive of the status quo. This move highlights the growing importance of TNCs in global EM matters and their wish to ensure that popular and official international initiatives do not adversely affect their interests. This coalition has been reinforced as a result of the Rio Conference in 1992, which called for increased TNC involvement in global environmental decision-making (Chatterjee & Finger 1994). Notable in this regard has been the creation of the Business Council on Sustainable Development. This organization was created in the run-up to Rio, when Stephan Schmidheiny, a leading Swiss businessman, persuaded a group of 48 corporate executives representing many of the world's leading TNCs to cooperate in the development of policies reflective of business EM interests. The immediate objective was to provide input to EM policy preparations for the Rio Conference (Schmidheiny 1992, Chatterjee & Finger 1994). This coalition has been a useful mechanism for the development of a "TNC perspective" on global environmental problems. The International Chamber of Commerce has been a comparable policy coalition in that it has sought to boost a pro-market set of EM interests. Through a "Business Charter for Sustainable Development", established in 1991, the coalition sought to build on the policy pronouncements of the WCED (1987) in such a way as to ensure that those interests were met (International Chamber of Commerce 1990, Eden 1994). These two initiatives have represented the efforts of one type of environmental manager to develop collectively a set of policies in keeping with the pursuit of financial predictability in EM.

The preceding discussion has highlighted the ambiguities surrounding the formation of policy coalitions, especially at the global level. These coalitions have become an increasingly important feature of modern EM (Young 1989).

At their best, they serve as a mechanism to bring together disparate interests in a common pursuit of predictability (e.g. CAP). The overall argument of this section has been to suggest that policy coalitions, however essential, are no panacea for the increasingly serious social and environmental problems confronting environmental managers.

There are considerable practical difficulties associated with the maintenance of trust among coalition partners in a context in which compromise and mutual sacrifice are essential. There is also the issue of potentially incompatible interests that are not necessarily dealt with by policy coalitions, which, as the TNC example illustrated, may be guided by highly partisan interests and concerns. Yet, as Chapter 5 suggested, the cooperation of all interested parties may be a prerequisite for developing a lasting consensus on EM issues.

Policy coalitions shift constantly in response to changing social and environmental problems. As this chapter has highlighted, they have yet to overcome the considerable differences that arise among environmental managers in multi-layered EM. Rapid changes in political, economic and cultural processes have nonetheless brought into question the ability of the state to promote predictability for all environmental managers in EM (Hurrell 1994). As such, it would appear that policy coalitions, increasingly transcending the boundaries of the nation state, will become increasingly important policy tools.

Conclusion

This chapter has explored the central importance of environmental policy-making in EM. It has described the policy-making characteristics of different types of environmental managers, and has suggested that policy coalitions are an important way in which they may seek to reconcile policy differences. To devise environmental policies may be the basis for enhancing predictability in EM, but it is precisely the central importance of policy-making to that endeavour that has often limited the ability of policy-makers to overcome fundamental political and economic differences.

What has emerged from Part III of this book is a fairly gloomy picture about the prospects for enhanced predictability in EM in the context of increasing social and environmental uncertainty (see Part II). The argument suggests that the problems confronting most environmental managers are too intractable to be solved by "technological fixes" – a common belief of many traditional accounts on EM. Rather, those problems may necessitate fundamental changes in the policies of all environmental managers. Such changes have implications for the ways in which EM is embedded in broader political and market processes. As Chapters 5 and 6 have highlighted, EM is much more than a technical process; rather, it is simultaneously a highly politicized and economically conditioned process in which multi-layered EM is linked in complex and often highly contradictory ways both to the market and to power structures at all scales. An understanding of the prospects for enhanced predictability in EM therefore facilitates understanding the political economy of multi-layered EM.

Does such an understanding mean that there can never be a sustainable EM based on a consensus embracing all environmental managers? A clear implication of the analysis so far is that enhanced predictability for some often means added uncertainty for others. Paradoxically, any conceivable set of policies that may lead to sustainable EM would appear to require fundamental changes in the political and economic organization of society, changes unlikely to occur without massive upheaval and further environmental degradation. However, new forms of human–environment interaction based on a syncretic mix of past and present EM practices might provide the basis for greater predictability in EM. This issue is addressed in Part IV.

PART IV

FUTURE DIRECTIONS IN ENVIRONMENTAL MANAGEMENT

Part III has explored whether political, market, and policy factors are conducive to the promotion of predictability in EM. It was argued that, at a general level, those factors have, if anything, added to social and environmental uncertainty in EM for most environmental managers. Yet, the dawning realization that not even the most powerful of environmental managers can escape the growing global environmental crisis may serve as the basis for a coordinated approach to EM problems in the future.

Part IV of this book takes up this issue by examining some of the different possibilities for EM in the twenty-first century. Although the discussion will inevitably be speculative in most respects, it is nonetheless vital to address the increasingly complex social and environmental issues that face environmental managers in multi-layered EM. In Chapter 8 the implications of technological change for the future EM practices of state and non-state environmental managers is considered. Chapter 9 summarizes the major arguments of the book. In reviewing the advantages of adopting an inclusive understanding of EM, the need for further research in order to elaborate the themes developed in this book is emphasized. As Chapter 1 noted, EM is not only a process but also a field of study. Therefore, Chapter 9 also considers the implications of a re-evaluation of EM for the nature and future direction of scholarly endeavour in this area.

CHAPTER 8
The future of environmental management

This book has focused on both the growing intensity of human impact on the environment and the increasing complexity of the ways in which humans have sought to promote predictability in EM. There has been a growing sense of pessimism about the prospects for humankind ever developing sustainable EM practices (Ehrlich & Ehrlich 1990, Lovelock 1995). The evidence provided in this book would tend, on balance, to corroborate this view.

There is nonetheless a range of social and technological changes and innovations taking place that may provide some grounds for optimism about the prospect for sustainable EM. On the one hand, recent grassroots initiatives around the world suggest that a partial devolution of power, consonant with local-level EM practices, might be one way in which to reduce uncertainty in EM (see Ch. 7). On the other hand, recent technological changes that encompass the development of new ways of understanding and constructing the environment through computers or biotechnology suggest possibilities for EM that may enable humankind to surmount the current global environmental crisis. Yet, such changes may also suggest a trend towards further concentration of EM decision-making in the hands of powerful political and economic elites, which may counteract advances achieved through local-level initiatives. What these diverse changes highlight, above all, is the complex interrelationship of political, economic, spatial, technological and ecological factors in the shaping of a future EM.

The following discussion presents a selective, and necessarily somewhat speculative, account of the possible future interactions between advanced technologies and EM. The point in doing so is not to "predict" the exact contours of a future EM, but rather to provoke thoughts about the potential impact of rapid technological changes on multi-layered EM.

Technology, empowerment and multi-layered environmental management

A prominent theme in contemporary literature is the issue of whether in future EM will need to take place through global or local environmental managers and mechanisms of control. This book has emphasized the importance of issues of scale and control in EM; Chapter 5, for example, focused on the role of differential power relations among environmental managers. These issues are becoming increasingly important in a context of growing social and environmental crises linked to the seemingly ever-increasing human impact on the environment.

The central issue in the "local versus global" debate is: which types of environmental managers are to have control over management practices that may be designed to promote sustainable EM? Put differently, the issue here is one of empowerment or disempowerment of diverse environmental managers in the future. The argument is linked to the related issue of blame in past unsustainable EM practices. It is suggested that those environmental managers who are largely to blame for previous or existing environmental degradation should not be entrusted with a substantial management role in the future. The point of contention here, of course, is over the apportioning of blame for previous malpractices.

One view, for example, is that much environmental degradation is linked to the rapid growth in population numbers and that sustainable EM is impossible in the face of continued population growth, especially in ELDCs (see Ch. 3). In this context, relying predominantly on local environmental managers who have contributed to this "population problem" is seen by some as unwise (Ehrlich & Ehrlich 1990, Swift 1993). Rather, trained "experts" in EM, notably linked to the state, business or international agencies, should take a lead role in promoting sustainable EM (Clark 1989, International Chamber of Commerce 1990).

In contrast, another view is that much of the blame for past and present EM malpractices must reside with precisely those "experts" linked to political and economic structures of power. Some argue that future EM should be based on a fundamental and far-reaching devolution of EM decision-making powers from global and national levels to the local level; that is, a transfer of decision-making away from the so-called "global environmental governance system" (Centre for Science and Environment 1992; see also Sachs 1993). At the local level, furthermore, it should be poor small-scale environmental managers (i.e. farmers, fishers, shifting cultivators) who ought to take the lead in promoting sustainable EM practices (Agarwal & Narain 1992, Chatterjee & Finger 1994, Pretty 1995). In this way, the "global versus local" debate mixes prescriptions for future EM based on the past and present practices of various environmental managers.

A central consideration in understanding this debate centres on the relationship between rapidly changing technologies and future opportunities for environmental managers. The relative opportunities and constraints facing environmental managers have certainly always been conditioned in part by technological change. Recent decades have nonetheless witnessed a process of accelerated technological change that may have far-reaching implications for EM, and that may lead to the transformation of human–environment interaction as we presently know it. In considering future EM, it is therefore essential to integrate discussions of scale and control with technological change. Of central concern in this regard is the extent to which technological change (i.e. computing, biotechnology) will empower certain environmental managers at the expense of other environmental managers. As Hill (1988) has observed, generally, technological change is never neutral but materially reinforces the power of selected groups in society (in a colonial context, see also Headrick 1981). Hence, how such change alters already highly unequal power relations between environmental managers warrants closer scrutiny.

In this regard, three aspects of the technology–EM interaction need to be highlighted (Fig. 8.1). First, the issue of "technological economies of scale" in EM must be considered. In effect, it would appear that certain technologies require a critical mass of financial and human resources and skills that is available only to more powerful environmental managers such as states and TNCs. Examples examined in detail in the next section include biotechnological changes linked to genetic manipulation, or planned efforts to "terraform" other planets (e.g. render Mars capable of human habitation). The general point here is that the ability to develop and apply such technologies is not avail-

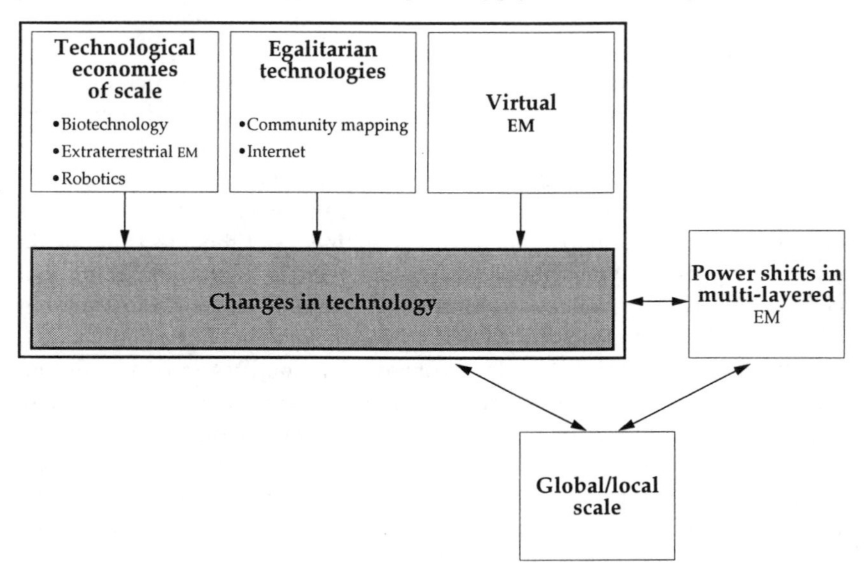

Figure 8.1 Technology, empowerment and multi-layered environmental management.

able to all environmental managers. This has potentially important implications in terms of the relative power of environmental managers. By definition, access is limited to these technologies – and, by extension, to the benefits that derive from the application of these technologies to EM. Hence, technological economies of scale will reinforce the power of certain environmental managers in relation to others through their ability to control the use of selected technologies.

In contrast, other technologies are potentially more "egalitarian" in their relationship to environmental managers. Such technologies may have the potential to enhance the EM practices and knowledge of most, if not all, environmental managers. As discussed on pp. 154–6, potential examples here relate to advances in computer technology (e.g. the Internet, computer-based mapping) that enable a potentially wide range of environmental managers to access services and knowledge. At first sight, the claim that computer applications represent "egalitarian" technologies might seem far-fetched; after all, it is such environmental managers as states and TNCs that have hitherto benefited disproportionately from applying such technology to their EM practices. There is growing evidence, nonetheless, that the diffusion of such technology to other environmental managers – including even small-scale farmers and hunter–gatherers – is accelerating, so that there is the possibility that all environmental managers (however poor and marginalized) may in the twenty-first century gain access to the benefits of this technology. In this manner, such "egalitarian" technologies have simultaneously the potential to enhance the EM capabilities of both large and small-scale environmental managers. At this stage, it is not clear how such technologies will affect the unequal power relations that characterize today's multi-layered EM.

Finally, technological change is transforming the ways in which the environment and EM are perceived by individuals. Specifically, the development of the so-called "virtual" technologies may have far-reaching implications for EM insofar as such technologies facilitate the creation of virtual environments, and hence the possibility for "virtual EM". The relationship between such virtual EM and "real-world" EM has yet to be defined, but the effects are again likely to be potentially enormous. For example, and as elaborated in the last section of this chapter, the spread of virtual EM capabilities may transform not only the ways in which environmental managers go about managing the real environment but may also affect levels of environmental use and consumption themselves. In effect, the distinction between imaginary and real EM may become increasingly blurred. In the process, virtual technologies may radically alter the terms of the local versus global debate insofar as virtual EM collapses space and time (i.e. through the ability to manipulate and re-create virtual environments). Depending on the extent of the link between virtual EM and real EM, and the diffusion of virtual technologies to environmental managers, such technological change may prove to be the most revolutionary of all in transforming existing power relations in multi-layered EM.

Technological economies of scale

The conventional view is that technological change serves to reinforce the position of more powerful environmental managers. For those technologies in which clear economies of scale exist in terms of access, use and control, this situation would appear to be the case. Perhaps the best-known example today of a technology that enhances the EM capabilities of powerful environmental managers is that of biotechnology. In recent decades a revolution has occurred in the field of scientific research and application of new biotechnologies. Specifically, scientists have developed means to manipulate and alter the genetic compositions of plant and animal species in order to enhance the ability of such species to produce greater quantities and quality of food and fibre products (cf. Table 2.1). As Kloppenburg (1988) and others have highlighted, such genetic manipulation is associated with the global capitalist economy insofar as most of the scientific research behind such manipulation is conducted either by large private firms or by state researchers, who are often beholden to those same firms through grant provisions. The giant TNC Monsanto, for example, routinely undertakes research contracts with university researchers on the principle that the research findings are treated as trade secrets and company property (ibid.).

Not only are the new biotechnologies linked directly to capital accumulation, they also serve to enhance the power of TNCs operating *vis-à-vis* other environmental managers. These TNCs have aggressively marketed new genetically manipulated seeds for agriculture and are providing fully integrated packages to farmers around the world; that is, seeds plus fertilizers plus herbicides/pesticides. In the future, these TNCs may develop these packages (e.g. to develop seeds capable of flourishing in degraded environments) in order to increase their hold over local-level environmental managers. In this way, TNCs can benefit from environmental degradation in that farmers will be dependent upon the biotechnologies those firms control for continuing production in degraded environments. In the process, the immense research and production costs that are associated with these evolving biotechnologies will ensure that in most cases only powerful environmental managers, such as TNCs, will have the necessary resources to benefit from, and control, those technologies.

Technological economies of scale will be most evident in what can be termed "extraterrestrial EM". In a context of growing anxiety over how to manage the Earth's environmental problems, it is easy to dismiss talk about extraterrestrial EM as pure science fiction. Not only is such EM already the subject of serious scientific investigation and experimentation, it is also likely to have major implications for the ways in which environmental managers seek to manage the environment on Earth. Indeed, space exploration has already served to transform human perceptions of the Earth in line with a new recognition of the Earth's fragility (Cosgrove 1994). Although this book is largely concerned with a terrestrially defined understanding of "environment" (see

Ch. 1), a more inclusive understanding – one that, for example, encompasses extraterrestrial space as "environment" – is appropriate here, given humankind's growing ability and desire to explore and exploit the resources of outer space.

To date, the primary contribution of humankind to extraterrestrial EM has been one of "environmental degradation" – notably the accumulation of "space junk" orbiting the Earth (Vogler 1995). Such extraterrestrial pollution mirrors the long history of human terrestrial environmental degradation examined in Chapter 3. Indeed, space is "the final frontier" and the term can be associated with pioneering mentalities. The perceived inexhaustibility of the extraterrestrial environment echoes earlier perceptions about selected terrestrial environments (e.g. North America, Amazonia; see Ch. 4).

The implications of extraterrestrial environmental degradation have yet to be fully absorbed. Nonetheless, not only does extraterrestrial "junk" pollute space near Earth, it also ultimately has direct implications for the terrestrial environment – that is, the uncontrolled re-entry of such waste to Earth. The case of the re-entry of the Soviet space station *Soyuz* 2 to Earth, and its swathe of destruction across Australia and Canada, is the most vivid example of negligent EM practices in space that impinge directly on the Earth itself.[1]

Human extraterrestrial EM practices have largely revolved around adding to the "environment". Other efforts seek to harness the perceived potential of other planets for future human use. An interesting example is the effort to develop EM techniques based on the idea of "terraforming" – that is, the manipulation of atmospheres on other planets to create conditions suitable for human residency and use. Initial ideas in this regard have focused on terraforming Mars, since that planet has, in theory, the requisite chemical ingredients necessary for such a transformation. A separate but related initiative is the much hyped (but also much maligned) Biosphere 2 experiment in the Arizona desert (USA). This project was initiated in 1983 and eight volunteer scientists ("Biospherians") began to inhabit this potentially self-contained life bubble in 1991 in order to test a human-created life-support system that would ultimately be capable of permitting colonization of other planets (Luke 1995).

As extraterrestrial EM moves from science fiction to science fact and EM reality, such EM will inevitably have implications for the relationship between environmental managers. Given the technological economies of scale, it is once again likely that financially and politically powerful environmental managers, such as states and TNCs, will be best placed to control access to this new and potentially powerful form of EM – and hence the political and economic benefits from such EM. States have played a central role in the preliminary experiments and research associated with extraterrestrial EM. Best known in this

1. The remnants of the Russian space station *Soyuz* 2 re-entered the Earth's atmosphere in an uncontrollable and unpredictable way. Some remnants (including parts of its nuclear reactor) crashed into uninhabited territory in Australia and Canada.

regard is the work of the National Aeronautics and Space Administration (NASA) – the agency responsible for the US government's space programme and for activities such as the *Apollo* programme, the *Voyager* mission and the Space Shuttle (Kerrod 1990, Cosgrove 1994). Other states have also been involved in similar programmes, including the former Soviet Union (e.g. *Soyuz* programme or *Mir* space station) or EU member states (*Ariane* programme, European Space Agency) (Bonnet & Manno 1994).

Given the high cost associated with these programmes, various states have found it convenient to cooperate in such efforts. The above-mentioned EU programme, but also the joint Russian–US–European initiative associated with a future space station, are typical of such interstate cooperation. States are not the only environmental managers with a stake in this process, since TNCs have also become involved in this potentially highly lucrative process. TNCs have paid for the privilege to use the Space Shuttle for various projects, such as the launching of telecommunication satellites or controlled environmental experiments under weightless conditions. The ability of this type of environmental manager to deploy substantial financial and human resources in the pursuit of diverse EM tasks enables it to undertake extraterrestrial EM. In contrast, other environmental managers such as farmers and hunter–gatherers are not able to participate in extraterrestrial EM (at least not in the foreseeable future).

In a similar fashion, there are technological economies of scale associated with the development and use of robots in EM. At present, science has developed artificial intelligence to a level sufficient to enable robots to undertake increasingly complex EM tasks. To date, such artificial intelligence has been applied to EM tasks that humans are unwilling, or unable, to undertake directly themselves. The most notable case in point is the use of robots to help in the management and disposal of hazardous waste. A good example is the use of robots in the decommissioning of nuclear facilities, or during accidents in nuclear power plants, the latter notably at Chernobyl in the Ukraine in 1986 (Mould 1992). As technological capabilities improve, nonetheless, the possibility of creating more complex human-like machines – increasingly referred to as "cyborgs" – raises the prospect of the growing substitution of robots for humans in an ever wider range of EM tasks. Indeed, once cyborgs become sophisticated enough to be able to actively and self-consciously manipulate the environment, it may be possible to speak of them as environmental managers themselves – environmental managers need not be human!

Whatever the prospects for robot-environmental managers, what seems clear at this stage is that there are clear technological economies of scale that favour the more powerful environmental managers, such as states and TNCs, at the expense of other environmental managers such as farmers and hunter–gatherers. Nevertheless, as the next section shows, the future of EM is also likely to be associated with what may become "egalitarian" technologies.

Egalitarian technologies

Some technological innovations seem to hold potential for widespread usage by a much broader range of types of environmental managers, and may be considered to be "egalitarian" technologies. The case of remote sensing and GIS technologies illustrates how technological change can result in the empowerment of a wide range of environmental managers. Such technologies originate in research and activities of powerful environmental managers such as the state and TNCs. Indeed, the ability to produce remotely sensed data is in part reliant on the launching of satellites which, as the previous section has highlighted, relates to technological economies of scale (Mack 1990). These powerful environmental managers have benefited from remote sensing and GIS technologies in terms of enhanced surveillance of human–environment interaction (e.g. monitoring of compliance by farmers with EU set-aside policies using remote sensing techniques). Nonetheless, the use of such data has not remained the exclusive preserve of either states or TNCs, but is becoming an increasingly important source of information for environmental NGOs, local-level environmental managers such as farmers, and even hunter–gatherers.

An important trend in the 1990s has been the growing reliance of local-level environmental managers on remotely sensed data in order to bolster arguments for the devolution of legal land and resource rights to the local level. Satellite data is used to define ancestral domains of indigenous and other local environmental managers in various parts of the world as part of campaigns by these people to assert local control over contested environments. On the islands of Luzon and Mindanao in the Philippines, for example, SPOT satellite imagery is combined with oral records of elders to map the land claims of the Lumad and Dumagat agriculturists, in response to the encroachment on the land of these indigenous environmental managers by lowland farmers and loggers (Braganza et al. 1993). In such local-level initiatives, critical input is typically provided by environmental NGOs, who use this positivist Western science and technology to help disadvantaged local communities against state-supported encroachments on their territorial resources.

In a similar manner, the growth of a globalized computer information community (based on the Internet), although initially centred on state institutions (i.e. universities and government departments), has become a more widely used tool for a range of state and non-state environmental managers (and many other users). It is certainly the case that, at least at present, Internet users are still disproportionately individuals linked to the state sector. For example, of the estimated 20000 plus documents linked to EM found on one search on the Internet in February 1996, a significant proportion of those sampled related to the policies and practices of governments and state institutions such as universities or research think-tanks. Most states now disseminate policy documents and conference proceedings on the Internet and are also using the Internet as a means to integrate EM activities. In Colombia, for example, the

Ministerio del Medio Ambiente (Ministry of Environment) is connecting its main field stations in nature reserves to the Internet (Mulligan 1996). Yet, such technology is also utilized by non-state environmental managers such as TNCs, international financial institutions and environmental NGOs. To take one example, the World Bank now provides full Internet facilities through the World Wide Web (WWW) for the dissemination of environmental information related to its projects (e.g. www.worldbank.org).

The use of the Internet by environmental NGOs is also a growing trend, acting often in association with local-level environmental managers. Environmental NGOs, in particular, are using this technology to obtain crucial data or to liaise with other environmental managers around the world in order to enhance their ability to help marginalized managers such as farmers and even hunter–gatherers. Transnational environmental NGOs such as Greenpeace and Friends of the Earth, and country-specific environmental NGOs such as the Environmental Research Division of the Manila Observatory (Philippines) and the Foundation for Ecological Recovery (Thailand), make use of the Internet as part of their various campaigns. For example, Friends of the Earth is using the WWW in order to publicize the polluting activities of business; Internet users in the UK can tap their local postcode into the computer in order to obtain FoE data, which details all the polluting companies operating within a 1.5km radius of that location (www.foe.co.uk/80/cri/html/[user's postcode]). One of the traditional weaknesses of local-level environmental managers has been a lack of awareness of wider political, economic and environmental issues and trends, as well as a lack of outside contacts (Esteva & Prakash 1992). With the help of environmental NGOs, this weakness is being overcome through the Internet, which enables them to have at their fingertips as wide a range of information as the state and TNCs.

To the extent that "knowledge is power", it can be said that the Internet has the potential to transform the interaction of different environmental managers radically. There are already more than 50 million Internet users worldwide. As its potential utility becomes more widely known, membership of this community across the globe is sure to grow, and in the process will incorporate hitherto marginalized environmental managers such as hunter–gatherers and farmers. It is as yet unclear what the ultimate impact of such an egalitarian technology will be on power relations among environmental managers. To the extent that the informational monopoly of states and TNCs is increasingly weakened, it can be seen nonetheless that some relatively weak environmental managers will gain strength *vis-à-vis* their hitherto predominant counterparts. In that regard, the Internet, as with remotely sensed data, may serve to weaken the power positions of states and TNCs with potentially enormous implications in terms of access to, and use of, environmental resources.

Indeed, the Internet is to information and knowledge as global financial markets are to economic activity, in that both enjoy a large measure of autonomy from state control. Try as they might, states are finding it increasingly difficult to control the global movement of money and information, and from an EM standpoint will find it ever more difficult to control and regulate access to information that is crucial to EM in the twenty-first century (cf. Hurrell 1994, Allen & Hamnett 1995). The battles of the future will, in part, revolve around the abilities of different environmental managers to make maximum use of egalitarian technologies in the promotion of their EM interests and goals.

It would appear that the development of egalitarian technologies is neutral in terms of the potential utility to, and empowerment of, a wide range of environmental managers. In this regard, such technologies stand in sharp contrast to those associated with economies of scale, in which only powerful environmental managers are able to participate in selected EM practices, and are able largely to control the benefits to be derived from those practices. Technological changes appear set to change EM practices in other ways that may transform the very notion of space and time as it has been conventionally understood in EM. Indeed, the new "virtual" computer technologies promise to blur the distinction between real and imaginary EM, so that the attitudes, perceptions and even "fantasies" of individual environmental managers become more important than ever before.

Towards a "virtual environmental management"?

One of the most intriguing technological developments in recent years has been that of so-called "virtual" technologies (cf. Doel & Clarke 1996, Light 1996). Such technologies involve the immersion of the individual in virtual worlds through the mechanism of computer-based visual stimulants. The possibilities for such technology range from enhanced leisure activities (i.e. computer games) to the creation of virtual operating theatres in order to aid medical practice. Although the link between virtual technologies and EM has yet to be explored systematically, the possibilities for a dramatic transformation in the policies and practices in multi-layered EM are immense.

In one respect, virtual technologies are likely to be significant as a tool available to a wide range of environmental managers. In this regard, such technologies are similar to GIS and the Internet, in that they have the potential to be egalitarian. What is particularly distinctive about the virtual technologies is that they challenge the boundaries between real and imaginary environments and EM. This raises the prospect of imaginary or virtual EM, which is the self-conscious manipulation of the environment – but in an imaginary world. This may have implications for real EM.

The link between imaginary environments and real world EM practices is not new. As Chapter 3 highlighted, Australian Aborigines sought to base their EM practices on an imagined world known as the "dreamtime"; however, new virtual technologies appear to have different EM implications from these "traditional" imaginings. Similarly, virtual technologies represent a significant elaboration of computer modelling and simulation techniques. The latter, although constituting a means to explore hypothetical EM scenarios, are nonetheless restricted in the sense that they are abstract (i.e. mathematical and symbolic) and do not necessarily convey a sense of environmental reality, as is the case with the latest virtual technologies (cf. Nemecek 1996). It is precisely the increasingly life-like qualities of the latter that are of potentially great significance for future EM.

Three key implications appear to be associated with a virtual EM. First, virtual technologies place a premium on the individual as a centre of the imaginary world (in contrast, for example, to Aboriginal dreamtime, which is based on a collectively constructed imaginary world). It is in the very nature of the technology – individual head sets and sensitized gloves used to manipulate the virtual environment and that individual's place in that environment – that the virtual reality is constructed and manipulated in light of an individual's actions. In this manner, the individual becomes the ultimate environmental manager – an omnipotent God-like being capable of creating and destroying imaginary worlds at the touch of a button. Indeed, all those entering the virtual world can become "environmental managers" in a sense. What is important here is not the accuracy with which virtual environments conform to real environments, but the importance attached to the virtual environment by an individual. As Tivers (1996: 6) has suggested with reference to computer games, "the degree to which [computer games] actually represent reality is of less interest than what they do actually represent to the player in the context of the game".

Secondly, in as much as the individual is at the heart of virtual EM, the attitudes, perceptions, and fantasies of the individual become critical in understanding the manipulation of imaginary worlds. But individuals will continue to be influenced by "opinion-formers" – for example, the influence of Gibson's (1988, 1993) conceptualization of "cyberspace" (Kneale 1996). Besides having far-reaching implications for theoretical debates over structure versus agency, the new virtual technologies will place an added premium on understanding the motivations and behaviour of individual environmental managers. It is far from clear how the increasing use of virtual technologies will alter individuals' attitudes and perceptions about EM generally. To the extent that all who so choose can become omnipotent "environmental managers", it will become important to evaluate the extent to which attitudes and perceptions shaped in the virtual world are then transferred to the real world (and vice versa). Indeed, the whole interaction between virtual and real EM may become a critical issue in a world increasingly influenced by virtual technologies.

Finally, and following from this, the ability of individuals – whether they be real-life environmental managers or other individuals – to construct and manage virtual environments may have severe implications in terms of actual use of the environment. For example, urban individuals, whose main connection with the environment may be for aesthetic enjoyment (e.g. visitors to national parks), may well visit real-world environmental attractions less often in light of the imaginary environments that they create and manage in their virtual worlds. Although these individuals are not necessarily real-world environmental managers, the fact that their impact on the environment may change because of their role as "virtual environmental managers" may have important implications for the policies and practices of environmental managers in the real world. In a similar vein, the opportunity for real-world environmental managers to simultaneously create virtual environments is likely to have substantial implications, not only in terms of their potentially modified EM practices, but also more generally in terms of their own attitudes and perceptions towards the environment and to EM as a concept and practice.

Unlike other technological changes, the advent of virtual technologies promises to change how environments and EM are perceived by both environmental managers and other individuals. It may be that virtual EM will develop to such an extent that current understandings of EM as a multi-layered process will have to be radically revised, particularly as the boundaries between real and virtual environments become ever more blurred (cf. Slouka 1995, Graham & Marvin 1996).

Conclusion

By how much will technological change materially transform the ability of environmental managers to pursue predictability in EM? As Part II of this book highlighted, technological change has had an ambiguous impact on human–environment interaction in the past – with the overall result being intensified use of the environment and increased uncertainty in EM. The current trajectory and pace of technological change would appear to hold equally ambiguous implications for the future of EM. On the one hand, certain technologies may enhance the power of already powerful environmental managers in multi-layered EM – thereby raising the prospect of intensified conflict. On the other hand, the possibility that so-called "egalitarian technologies" may strengthen the positions of hitherto weak environmental managers, *vis-à-vis* their more powerful counterparts, may herald a new era in which the defining characteristics are comprehensive cooperation in EM based on mutual respect and reciprocity. And the prospect of growth in virtual EM even brings into question the very nature of multi-layered EM in a future world in which real and imaginary environments uneasily coexist.

Technological change promises to have as dramatic an impact on future EM, as it has in the past. The growing importance of technological change in human–environment interaction (including extraterrestrial environments) does not necessarily lead to the rosy scenarios imagined by such technocentric thinkers as Simon & Kahn (1984). According to these writers, technology will "fix" the world's environmental problems. Such an optimistic worldview flies in the face of the highly unequal and politicized nature of multi-layered EM. Indeed, the argument of Part III of this book was that it was in the nature of multi-layered EM to involve conflict between environmental managers. In other words, uncertainty appeared to be a key feature of EM.

In contrast, this chapter has suggested that rapid technological change may require a revised assessment of the prospects for sustainable EM based on comprehensive cooperation among different types of environmental managers. On the one hand, such change (i.e. technological economies of scale) is liable to reinforce the power of already-powerful environmental managers (e.g. the state, TNCs), leading those environmental managers to believe that there is no need for comprehensive cooperation in the face of uncertainty in EM. In other words, such environmental managers may see technological change as an opportunity to continue EM policies and practices based on the transference of uncertainty to weaker environmental managers. In this regard, technological change may alter the capabilities of environmental managers, although leaving unmodified their self-interested goal-seeking behaviour.

On the other hand, to the extent that "egalitarian" technologies live up to their name, some of the advantages of more powerful managers may be eroded as even relatively weak managers (e.g. farmers, hunter–gatherers) gain access to potentially powerful computer-based technologies. Under such circumstances, traditionally more powerful managers may acknowledge the necessity of adopting an EM approach based on cooperation among all environmental managers. In this manner, egalitarian technologies may reinforce the fledgling efforts towards cooperative EM examined in Chapters 5 and 7.

Whereas it is clear that technological change will continue to have an important impact on the ability of environmental managers to pursue predictability in their EM practices, it remains far from clear how the various types of technological change will affect the constraints and opportunities facing all environmental managers. What is clear is that such change underscores the need to take into account an inclusive approach to the process of EM as humankind approaches a new millennium. The next and concluding chapter of the book considers the broader implications of this more inclusive understanding of the process of EM. It suggests areas for further research and, building on Chapter 1, considers how the field of EM itself is to be understood.

CHAPTER 9
Conclusion

The argument of this book has been that a more inclusive understanding of EM is necessary, if the full range of management issues associated with human–environment interaction is to be appreciated. A re-evaluated EM needs to investigate the policies and practices of state, as well as non-state, environmental managers. What remains to be considered are the implications of this reassessment for EM as a field of study. Following a brief summary of the principal arguments of this book, the implications for further research and for the field itself are assessed.

Environmental management as a multi-layered process

The initial premise of the book was that many traditional accounts of EM have failed to capture the full breadth of EM as a process. Many of these accounts have been little more than "technical" manuals designed to provide policy-relevant advice to environmental managers linked to the state. They have been essentially prescriptions of "how to" manage specific components of the Earth's environment. Such accounts have a surreal quality about them insofar as they often are disconnected from the political, economic and cultural contexts in which environmental problems occur and that need to be taken into account in the decision-making process. In contrast, this book builds on an emerging literature that seeks to locate EM at the heart of social science research (see below). The goal is to understand precisely those connections between social context and EM practices that are critical to any full appreciation of contemporary EM.

Central to this endeavour has been the development of an analytical framework based on the notion of EM as a multi-layered process (see Fig. 1.1). This framework has several advantages. First, and by including state and certain non-state actors as environmental managers, it facilitates a recognition of the EM capabilities of the latter. The book, therefore, has been concerned as much to analyze, say, the policies and practices of TNCs, farmers and hunter–gatherers as it has been involved with appreciating the role of the state (and its constituent agencies) in multi-layered EM. Indeed, by suggesting that the state is only one type of environmental manager among many others, the book highlights those qualities possessed by the state that potentially distinguish it from other non-state environmental managers.

Secondly, it follows from understanding EM as a multi-layered process that there needs to be a continuing revision of the ways in which non-state environmental managers in particular are understood. In traditional accounts, they are often not seen as environmental managers at all, but rather as mere recipients of the advice, "expertise" or even orders of the "real" (i.e. state) environmental managers. In contrast, in the re-evaluated EM explored in this book, non-state environmental managers are often key players. Hence, they need to be accorded a prominent place in any account that seeks to understand the management dimensions to human–environment interaction.

Thirdly, conceiving of EM as a multi-layered process also underscores the need to appreciate how the EM policies and practices of diverse environmental managers are influenced by political, market and sociocultural factors. Environmental managers relate to each other in complex ways, reflecting power relations and market positions. Simultaneously, they also devise policies about EM that reflect not only political and economic constraints and opportunities, but also attitudes towards the environment linked to broader cultural worldviews and discourses. By emphasizing these interconnections, this book has, above all, sought to understand EM as a social rather than as a technical process.

In order to understand how state and non-state environmental managers attempt to manage the environment, the book has explored the utility of two interlinked concepts: predictability and uncertainty. It was argued that the central goal of all environmental managers is to enhance predictability in EM. Such predictability has environmental, sociocultural, political and financial dimensions, and different environmental managers pursue these dimensions in a variety of ways through their EM policies and practices (see Table 2.2). Yet, to understand the constraints and opportunities that influence the ability of any environmental manager to pursue predictability, it was suggested in Part III that political, market and policy processes needed to be appreciated. One important insight was that increased predictability for some environmental managers may result in increased uncertainty for other environmental managers. A further insight was that increased predictability for any environmental manager is not synonymous with sustainable EM. Predictability is, thus, a multi-faceted concept that helps to clarify the rationale for EM.

The quest for predictability is, above all, a response to uncertainty in EM. Such uncertainty has environmental, sociocultural, political and market dimensions, which in complex ways constrain the activities of environmental managers. Environmental managers are differentially affected by the different dimensions of uncertainty, depending on specific patterns of social and environmental interaction. This book has also highlighted how the EM practices of state and non-state environmental managers have resulted in an overall increase in uncertainty in EM over time. Uncertainty is certainly linked, in part, to the vicissitudes of ecological processes. It is nonetheless also associated with the increasing intensity of human–environment interaction. Further, more uncertainty has arisen as different types of environmental managers come into increasing conflict over management of the environment at local, national, and global scales. Indeed, as the emergence of TNCs and environmental NGOs as environmental managers in recent decades illustrates, the multi-layered nature of EM is becoming more complex – a process set to intensify in the twenty-first century because of rapid technological changes that may be transforming the ways that people perceive and utilize the environment.

Overall, the general tone of this book has been rather pessimistic about the prospects for environmental managers to enhance predictability in EM. In a sense, such pessimism is associated with the very act of acknowledging that EM is a multi-layered process. This more inclusive understanding underscores the complexities and difficulties associated with the attainment of comprehensive cooperation among environmental managers in the pursuit of predictability. The point of this book, though, is not to deny the possibility of such cooperation (indeed, it suggested tentative efforts towards that end) but to acknowledge fully and explore the obstacles to the pursuit of predictability in EM. Therefore, this book promotes an understanding of the process of EM conforming to contemporary conditions under which environmental managers operate.

The analysis contained in this book also raises issues that go beyond EM as a process. As was hinted at in Chapter 1 already, it also has far-reaching implications for how EM may be understood as a field of study.

Rethinking environmental management as a field of study

Re-evaluating EM as a process necessitates that the understanding of the field of EM itself needs to be revised. This revision includes questioning the subjects researched, the beneficiaries of this research, and indeed the identity of the researchers themselves. In other words, if a central objective in understanding EM is to promote a more inclusive appreciation of EM as a process, then efforts to rethink EM as a field of study must be concerned principally with "decentring" the state from its hitherto privileged position in EM as a field of study.

In this regard, what is suggested here for EM is akin to efforts in many other research fields that seek to go beyond the state in order to capture more effectively the complexity of human–environment interaction. A case in point is current research in the field of international relations, a subdiscipline of political studies, which challenges the centrality of the state in the analysis of international political problems (Walker & Mendlovitz 1990). Indeed, the growth of a whole host of political problems associated with global environmental change is a main impetus behind this movement (Lipschutz & Conca 1993, Vogler & Imber 1996). The argument presented in this book forms part of a broader effort to incorporate the concerns and interests of non-state actors into research.

A necessary element in the effort to go beyond a state-centric EM is a re-evaluation of the nature and purpose of EM as a field of study. It is important to reiterate in this regard the origins and development of EM as a policy-relevant field of study geared towards advising the state on managing the environment (MacNeill 1971, Dorney 1987). EM has developed as a community of scholars and practitioners who draw upon a diverse range of technical skills – such as planning, law, ecology and biology – in order to help state environmental managers in problem-solving exercises (see Fig. 1.2). EM has been a field of study largely dominated by "experts" whose main rationale has been to provide advice to the state on environmental issues. This traditional approach tends to neglect those political, economic and cultural issues that constrain and enable all types of environmental managers. Just as EM is not simply a technical process, so too the community of scholars who contribute to the understanding of this field of study ought not to be composed solely of technicians.

Therefore, it is essential that, if EM is to become a field of study relevant to the world's pressing environmental problems, it must draw upon a much wider array of expertise than has hitherto been the case. Although a place for policy advisers and scientific experts linked to the state will undoubtedly remain, more room will also be necessary for other individuals hitherto not represented in this community of scholars. In this regard, researchers linked to environmental NGOs, TNCs, and international financial institutions will need to be increasingly incorporated if insights into the policies and practices of these types of environmental managers are to be taken on board. Perhaps more controversial is the assertion that activists representing farming, fishing and hunter–gatherer communities should be seen to have a vital input to the development of the field of EM.

To a certain extent, the process of broadening the membership of the scholarly community is already under way. For example, the work of such activist scholars as Agarwal & Narain (1992), Khasiani (1992) and Sachs (1993) represents not only a fundamental challenge to the principles of traditional EM, but also a vital contribution to efforts to base EM on a more inclusive understanding of the subject matter. Such scholars are initiating important changes to the field of EM by virtue of their very activities.

In the measure that EM becomes a more inclusive scholarly community – based on, in turn, a wider acknowledgement of EM expertise – the subjects habitually studied by scholars will also inevitably change. Indeed, this book contributes to a growing literature that calls for the development of a more inclusive understanding of what is considered to be important to research in EM. At one level, as this book has highlighted, this will entail an appreciation of the policies and practices of a wide range of types of environmental managers. At another level, by adopting a more inclusive understanding of the subject matter, scholars in the field of EM will also need to rethink the methodologies they employ in undertaking their research. Reflecting the scientific training of traditional scholars, the field has been largely characterized by a highly quantitative and positivist scientific approach. As this book has emphasized, such an approach is only one way of understanding the environment and EM. As a result, there is a need to draw upon hitherto neglected qualitative methodologies in research, such as participant observation, in-depth interviews, and even the use of oral histories (Braganza et al. 1993). This is not to abandon quantitative techniques; rather, it is to say that non-positivist research techniques are an essential part of a re-evaluated EM.

Just as the subject (and personnel) in EM will need to change, so too will the intended audience for the research undertaken have to be adapted in order to develop a more inclusive approach to the subject matter. Traditionally, scholars have undertaken research with the main intention of providing a product usable by the state. That the intended audience for research in EM has hitherto been almost exclusively the state is not surprising, given that many scholars have derived the funds needed to undertake research directly from the state. In effect, they have been consultants to the state on EM issues. Once it is recognized that states are by no means the only type of environmental manager, it becomes important for scholars to address their research to a much wider audience. This will have implications in terms of the sources of funding for future research – primarily entailing a reduction in the importance of the state as a funding agency in favour of a more diverse range of sponsors, including TNCs, environmental NGOs and farmers' organizations. Indeed, this process is already occurring as scholars draw upon organizations such as Friends of the Earth or Shell Oil in the funding of their research, and in the process provide innovative new insights into EM processes (e.g. Friends of the Earth 1994a).

More importantly, there will also be a need to ensure that the dissemination of knowledge encompasses the range of environmental managers in multi-layered EM. Once again, this process is already under way. For example, the research on upland EM practices in the Philippines noted in Chapter 8 is disseminated to officials in the Department of Environment and Natural Resources in the Philippines, but above all is provided to the environmental managers at the heart of the research to enable those individuals to assert their EM rights *vis-à-vis* other environmental managers (Braganza et al. 1993). In effect, this process reflects a wider set of developments in social science research, designed to ensure that hitherto marginalized individuals or groups gain access to the fruits of scholarly research (Chambers 1983, Chambers et al. 1993, Pretty 1995).

Therefore, the utility of a re-evaluated EM is to ensure that the field is based on an inclusive appreciation of the role and purpose of EM in a rapidly changing world. Indeed, such an appreciation is essential if EM is to embark on new directions in the twenty-first century that will be designed to render it capable of responding to the intellectual and practical challenges that undoubtedly lie ahead.

In turn, such an appreciation has important implications for how EM relates to geography and other cognate disciplines. As Chapter 1 has highlighted (see Fig. 1.2), EM is characterized by a potentially wide range of disciplinary influences. Insofar as EM is to be understood as a multi-layered process, this will affect the relative importance of those different disciplinary influences in the further development of EM as a field of study. The more inclusive understanding of EM explored in this book would suggest first of all the need to emphasize the importance of selected disciplines hitherto at the margin of EM. For example, the importance attached to both politics and attitudes (culture) in appreciating multi-layered EM highlights the importance of drawing upon theoretical and empirical developments in the fields of politics and anthropology (especially key subfields such as environmental politics and cultural ecology). Such an approach stands in contrast to many traditional accounts in which scholars have drawn primarily upon work in policy and planning, management studies, resource management and environmental science. The point here is not to suggest that these "traditional" disciplinary sources of ideas and information have no place in a re-evaluated EM. Rather, it is to underscore the necessity of drawing more extensively upon work in other disciplines in line with the more inclusive understanding of the subject matter explored in this book. Just as the subject matter, methodologies and intended audience will need to be adjusted, so too the specific disciplinary mix of influences will also inevitably require modification if EM is to be relevant to the environmental issues and concerns in the twenty-first century.

It follows from the argument of this book that EM ought to be primarily orientated towards the social sciences. The disciplines from which it may draw inspiration will be predominantly located in this sector (see Fig. 1.2). In this regard, this book's analysis of EM as a multi-layered process emphasizes the central importance of understanding how the interaction of environmental managers – and the environmental attitudes, worldviews and discourses of those managers – conditions human use of the environment. Such a quest is largely based in the social sciences.

In this respect, the contrast between EM and environmental science needs reiteration. Whereas the latter is predominantly based in the natural sciences, the former is largely situated in the social sciences. That is not to say, that each field of study does not impinge on topics habitually located in the other field of study's domain. Hence, research in environmental science suggests how the understanding of physical and ecological processes has a bearing on EM considerations. In a similar fashion, EM is concerned in part to clarify how the interaction of environmental managers in multi-layered EM has social and ecological implications.

O'Riordan (1995a) argued the case for a policy-relevant environmental science. Indeed, he has highlighted at a general level the relationship between environmental science and EM. Nothing in the argument of this book would contest the need for such a relationship in the future. That said, a central implication of a re-evaluated EM would be a recognition that these two fields of study, although linked by a common concern with human–environment interaction, are nonetheless distinct fields of enquiry that often require different research methodologies or theoretical frames of reference. Whereas environmental science and EM usefully may draw upon research from each other, it is important to be clear of the different objectives and purposes of each field of study.

In contrast, a re-evaluated EM is likely to continue to draw extensively upon the discipline of geography. Throughout this book, the importance of space and scale issues has been highlighted, suggesting that EM could profit from the ways in which human geography, in particular, addresses similar issues. The linkage here would be based on appreciating the theoretical and empirical developments in human geography that would help to clarify the operation of multi-layered EM. For example, work by geographers into the ways in which a technologically sophisticated and globalized capitalist system has served to collapse space and time (e.g. global financial flows) could provide useful insights into how contemporary technological and economic change has a differential impact on environmental managers operating in multi-layered EM (Harvey 1989, Murdoch & Marsden 1995).

In this manner, a reassessed EM will need to draw upon work in human geography, as well as other social sciences, to elaborate upon the distinctive rationale explained in this book. At present, EM can be seen as a field of study associated predominantly with the discipline of geography, yet crucially drawing upon other disciplinary influences (see Fig. 1.2). As environmental problems at all scales intensify, the central importance of understanding those problems will grow – especially in light of the concerns of a re-evaluated EM. In the process, the future may witness the progression of EM from a field of study into a discipline in its own right. Such a progression would constitute a paradigm shift of potentially great intellectual and practical significance for how EM is understood and implemented.

Conclusion

In a sense, this book has arisen from a paradox. Human perceptions about the environment have undergone dramatic changes over a long period of time. However, human understanding about EM as a process, and as a field of study, has remained relatively static. Indeed, EM as a field of study may have become less relevant to the main issues associated with human–environment interaction, and may be in jeopardy of becoming marginalized as a form of intellectual enquiry.

The objective of this book has been to explore the possible bases of a more inclusive understanding of EM as a multi-layered process that might counter such marginalization. The book has done so primarily by suggesting the utility of an analytical framework in which the activities of environmental managers are understood in relation to the concepts of uncertainty and predictability. The emphasis has been on general analytical concerns rather than empirical explanation.

An important task for future research is to explore further the applicability of this analytical framework in terms of the specific activities of a wide range of environmental managers. Such work, in particular, might usefully consider the extent to which the environmental policies of state and non-state environmental managers translate readily into practice or not, thereby exploring potential tensions between the evolution of policy and the course of day-to-day practices.

Indeed, the gap between rhetoric and reality in state EM policies is already becoming increasingly apparent in the aftermath of the Rio Earth Summit 1992. The general difficulties surrounding the pursuit of predictability in a context of mounting social and environmental uncertainty would suggest that this gap between policy and practice may also extend to many, if not all, non-state environmental managers. The tensions surrounding such discrepancies are becoming increasingly evident in the growing popular and scholarly confusion concerning the meaning of "sustainable development". As this book has emphasized, sustainability has not always been a preoccupation of environmental managers, and there is little evidence to suggest that this situation will be altered in the twenty-first century.

It is all the more important, therefore, to attempt to locate EM within a framework that acknowledges the various aims and interests of environmental managers as they interact with each other and with the environment. Although certainly not providing a comprehensive treatment of all the issues related to such a framework, this book has nonetheless sought to identify the possible elements that need to be incorporated in such a re-evaluated EM.

REFERENCES

Abel, N. O. J. & P. Blaikie 1989. Land degradation, stocking rates and conservation policies in the communal rangelands of Botswana and Zimbabwe. *Land Degradation and Rehabilitation* **1**, 101–123.

Adams, J. 1995. *Risk*. London: UCL Press.

Adams, W. M. 1990. *Green development – environment and sustainability in the Third World*. London: Routledge.

Adas, M. 1974. *The Burma Delta: economic development and social change on an Asian rice frontier, 1852–1941*. Madison, Wisconsin: University of Wisconsin Press.

—1981. From avoidance to confrontation: peasant protest in pre-colonial and colonial South East Asia. *Comparative Studies in Society and History* **23**, 217–47.

—1989. *Machines as measures of men – science, technology, and ideologies of Western dominance*. Ithaca, New York: Cornell University Press.

Agarwal, A. & S. Narain 1989. *Towards green villages*. New Delhi: Centre for Science and the Environment.

—1992. *Towards a green world*. New Delhi: Centre for Science and Environment.

Ajzen, I. 1988. *Attitudes, personality, and behavior*. Chicago: Dorsey Press.

—1991. The theory of planned behaviour. *Organizational Behavior and Human Decision Processes* **50**, 179–211.

Ajzen, I. & M. Fishbein 1980. *Understanding attitudes and predicting social behavior*. Englewood Cliffs, New Jersey: Prentice Hall.

Alexander, D. 1985. Introductory remarks from the new editor in chief. *Environmental Management* **9**(6), 461.

Allen, J. & C. Hamnett (eds) 1995. *A shrinking world? Global unevenness and inequality*. Oxford: Oxford University Press.

Allison, L. 1974. *Environmental planning: a political and philosophical analysis*. London: George Allen & Unwin.

Atchia, M. & S. Tropp (eds) 1995. *Environmental management: issues and solutions*. Chichester: John Wiley.

Baas, S. 1993. Range capacity and range condition in the Central Rangelands of Somalia. In *Factors limiting pastoral production in Central Somalia*, H. J. Schwartz, J. Janzen, M. Baumann (eds), 83–99. Bonn: GTZ Schriftenreihe.

Bailey, S. 1995. The effectiveness of conservation development projects in reducing the dependency of communities living in or around national parks in Ecuador on forest resources contained within the parks. Paper presented at the Interdisciplinary Research Network of Environment and Society (IRNES), September 1995, Keele University.

Baldock, D., C. Cox, P. Lowe, M. Winter. 1990. Environmentally sensitive areas: incrementalism or reform? *Journal of Rural Studies* **6**(2), 143–62.

Banks, G. 1993. Mining multinationals and developing countries: theory and practice in Papua New Guinea. *Applied Geography* **13**, 313–27.

Barbier, E. B. 1987. The concept of sustainable economic development. *Environmental Conservation* **14**, 101–110.

—1989. *Economics, natural-resource scarcity and development*. London: Earthscan.

—1993. Economic aspects of tropical deforestation in Southeast Asia. *Global Ecology and Biology Letters* **3**(4–6), 215–34.

Barbier, E. B., J. C. Burgess, T. M. Swanson, D. W. Pearce 1990. *Elephants, economics and ivory*. London: Earthscan.
Barnes, T. & J. Duncan (eds) 1992. *Writing worlds: discourse, text and metaphor in the representation of landscape*. London: Routledge.
Barney, G. O. (study director) 1980. *The Global 2000 report to the President*. Washington DC: US Government Printing Office.
Bassett, T. J. 1988. The political ecology of peasant–herder conflicts in the northern Ivory Coast. *Annals of the Association of American Geographers* **78**, 453–72.
Beck, U. 1992. *Risk society*. London: Sage.
Beckerman, W. 1974. *In defence of economic growth*. London: Jonathan Cape.
—1995. *Small is stupid: blowing the whistle on the greens*. London: Duckworth.
Beinart, W. & P. Coates 1995. *Environment and history – the taming of nature in the USA and South Africa*. London: Routledge.
Benedick, R. E. 1991. *Ozone diplomacy: new directions in safeguarding the planet*. Cambridge, Massachusetts: Harvard University Press.
Benton, T. 1993. *Natural relations: ecology, animal rights and social justice*. London: Verso.
—1994. Biology and social theory in the environment debate. In *Social theory and the global environment*, M. Redclift & T. Benton (eds), 28–50. London: Routledge.
Benton, T. & M. Redclift 1994. Introduction. In *Social theory and the global environment*, M. Redclift & T. Benton (eds), 1–27. London: Routledge.
Berry, B. J. L. 1990. Urbanization. In *The Earth as transformed by human action – global and regional changes in the biosphere over the past 300 years*, B. L. Turner, W. C. Clark, R. W. Kates, J. F. Richards, J. T. Mathews, W. B. Meyer (eds), 103–120. Cambridge: Cambridge University Press.
Berry, S. 1989. Social institutions and access to resources in Africa. *Africa* **59**, 41–55.
Birch, J. W. 1973. Geography and resource management. *Journal of Environmental Management* **1**, 3–11.
Black, J. 1970. *The dominion of man: the search for ecological responsibility*. Edinburgh: Edinburgh University Press.
Blackburn, T. & K. Anderson (eds) 1993. *Before the wilderness: environmental management by native Californians*. Menlo Park, California: Ballena Press.
Blaikie, P. 1985. *The political economy of soil erosion in developing countries*. London: Longman.
Blaikie, P. & H. Brookfield 1987. *Land degradation and society*. London: Methuen.
Blaikie, P., T. Cannon, I. Davis, B. Wisner 1994. *At risk – natural hazards, people's vulnerability, and disasters*. London: Routledge.
Blowers, A. (ed.) 1993. *Planning for a sustainable environment*. London: Town and Country Planning Association.
Bluehdorn, I. 1995. Environment NGOs and "new politics". *Environmental Politics* **4**(2), 328–32.
Boehmer-Christiansen, S. 1994a. Politics and environmental management. *Journal of Environmental Planning and Management* **37**(1), 69–85.
—1994b. The precautionary principle in Germany – enabling government. In *Interpreting the precautionary principle*, T. O'Riordan & J. Cameron (eds), 31–61. London: Earthscan.
Bonnet, R. M. & V. Manno 1994. *International cooperation in space: the example of the European Space Agency*. Cambridge, Massachusetts: Harvard University Press.
Boomgard, P. 1994. Sacred trees and haunted forests – Indonesia, particularly Java, 19th and 20th centuries. In *Asian perceptions of nature: a critical approach*, O. Bruun & A. Kalland (eds), 39–53. London: Curzon Press.
Born, S. M. & W. C. Sonzogni 1995. Integrated environmental management: strengthening the conceptualization. *Environmental Management* **19**(2), 167–82.

Boserup, E. 1993 (1965). *The conditions of agricultural growth – the economics of agrarian change under population pressure*. London: Earthscan.

Botkin, D. & E. Keller 1995. *Environmental science: Earth as a living planet*. Chichester: John Wiley.

Bowlby, S. R. & M. S. Lowe 1992. Environmental and green movements. In *Environmental issues in the 1990s*, A. Mannion & S. R. Bowlby (eds), 161–74. Chichester: John Wiley.

Bowler, I., C. Bryant, D. Nellis (eds) 1992. *Contemporary rural systems in transition I: agriculture and environment*. Wallingford: CAB International.

Bowonder, B. 1986. Environmental management problems in India. *Environmental Management* **10**(5), 599–610.

—1987. Integrating perspectives in environmental management. *Environmental Management* **11**(3), 305–316.

Braganza, G., J. B. Ong, G. J. Tengco, E. Wijanco 1993. *Upland Philippine communities: guardians of the final forest frontiers*. Research Network Report 4, Centre for Southeast Asia Studies, University of California, Berkeley.

Braganza, G., J. Ong, C. Vicente 1994. *Upland Philippine communities: securing cultural and environmental stability*. Manila: Environmental Research Division.

Bramble, B. J. & G. Porter 1992. Non-governmental organizations and the making of US international environmental policy. In *The international politics of the environment*, A. Hurrell & B. Kingsbury (eds), 313–53. Oxford: Oxford University Press.

Brenton, T. 1994. *The greening of Machiavelli: the evolution of international environmental politics*. London: Earthscan.

Breyman, S. 1993. Knowledge as power: ecology movements and global environmental problems. In *The state and social power in global environmental politics*, R. D. Lipschutz & K. Conca (eds), 124–57. New York: Columbia University Press.

Briggs, D. & F. Courtney 1989. *Agriculture and environment*. Harlow: Longman.

Brimblecombe, P. 1987. *The big smoke*. London: Methuen.

Britton, S., R. LeHeron, E. Pawson (eds) 1992. *Changing places in New Zealand*. Christchurch: New Zealand Geographical Society.

Broad, R. 1993. *Plundering paradise: the struggle for the environment in the Philippines*. Berkeley: University of California Press.

Broad, R. 1994. The poor and the environment: friends or foes? *World Development* **22**(4), 811–22.

Bromley, D. W. (ed.) 1995. *The handbook of environmental economics*. Oxford: Blackwell.

Brookfield, H. C. 1988. The new great age of clearance and beyond. In *People of the tropical rain forest*, J. S. Denslow & C. Padoch (eds), 205–224. Berkeley: University of California Press.

Brophy, M. 1996. Environmental politics. In *Corporate environmental management*, R. Welford (ed.), 92–103. London: Earthscan.

Brown, K. & D. Pearce (eds) 1994. *The causes of tropical deforestation*. London: UCL Press.

Brown, L. R. & H. Kane 1994. *Full house: assessing the Earth's population carrying capacity*. New York: Norton.

Brown, V., D. I. Smith, R. Wiseman, J. Handmer 1995. *Risks and opportunities: managing environmental conflict and change*. London: Earthscan.

Browne, W., J. R. Skees, L. E. Swanson, P. B. Thompson, L. J. Unnevehr 1992. *Sacred cows and hot potatoes: agrarian myths in agricultural policy*. Boulder, Colorado: Westview.

Bruun, O. & A. Kalland (eds) 1994. *Asian perceptions of nature: a critical approach*. London: Curzon Press.

Bryant, R. L. 1992. Political ecology: an emerging research agenda in Third World studies. *Political Geography* **11**, 12–36.

—1994a. Shifting the cultivator: the politics of teak regeneration in colonial Burma. *Modern Asian Studies* **28**, 225–50.

—1994b. From laissez-faire to scientific forestry: forest management in early colonial Burma 1826–85. *Forest and Conservation History* **38**, 160–70.

—1996. Romancing colonial forestry: the discourse of "forestry as progress" in British Burma. *Geographical Journal* **162**(2), 169–78.

—1997. *The political ecology of forestry in Burma, 1824–1994*. London: Hurst.

Bryant, R. L. & S. Bailey 1997. *Third World political ecology*. London: Routledge.

Buchanan, K. 1973. The White north and the population explosion. *Antipode* **5**(3), 7–15.

Buckley, R. 1991. *Perspectives in environmental management*. Berlin: Springer.

Bullard, R. D. (ed.) 1993. *Confronting environmental racism: voices from the grassroots*. Boston: South End Press.

Buller, H. J. 1992. Agricultural change and the environment in western Europe. In *Agricultural change, environment and economy*, K. Hoggart (ed.), 68–88. London: Cassell (Mansell).

Burrows, P. 1979. *The economic theory of pollution control*. Oxford: Martin Robertson.

Buttel, F. H. 1986. Sociology and the environment: the winding road toward human ecology. *International Social Science Journal* **38**(3), 337–56.

Caldwell, L. K. 1990. *International environmental policy: emergence and dimension*, 2nd edn. Durham, North Carolina: Duke University Press.

Callicot, J. B. & R. T. Ames 1989. *Nature in Asian traditions of thought*. Albany, New York: SUNY Press.

Carr, S. & J. Tait 1991. Differences in the attitudes of farmers and conservationists and their implications. *Journal of Environmental Management* **32**, 281–94.

Carson, R. 1962. *Silent spring*. Boston: Houghton Mifflin.

Castillon, D. A. 1992. *Conservation of natural resources: a resource management approach*. Dubuque, Iowa: Brown.

Centre for Science and Environment 1992. Statement of the Centre for Science and Environment on global environmental democracy. *Alternatives* **17**, 261–79.

Chambers, R. 1983. *Rural development: putting the last first*. London: Longman.

—1987. *Sustainable livelihoods, environment and development: putting poor rural people first*. Discussion Paper 240, Institute of Development Studies, University of Sussex.

Chambers, R., A. Pacey, L. A. Thrupp (eds) 1993. *Farmer first: farmer innovation and agricultural research*. London: Intermediate Technology.

Chatterjee, P. & M. Finger 1994. *The earth brokers: power, politics and world development*. London: Routledge.

Chia Lin Sien 1987. *Environmental management in Southeast Asia*. Syracuse: New York University State Press.

Choi, Y. H. (ed.) 1985. Culture and the environment. *Environmental Management* [special issue] **9**(2), 95–174.

Clark, W. C. 1989. Managing planet earth. *Scientific American* **261**(3), 18–26.

Claro, E. & G. A. Wilson 1996. Trans-Pacific wood chip exports: the rise of Chile. *Australian Geographical Studies* **34**(2), 185–99.

Clayton, K. 1995. The threat of global warming. In *Environmental science for environmental management*, T. O'Riordan (ed.), 110–30. Harlow: Longman.

Cocklin, C., S. Parker, J. Hay 1992. Notes on cumulative environmental change, I: concepts and issues. *Journal of Environmental Management* **35**(1), 31–50.

Colby, M. E. 1991. Environmental management in development: the evolution of paradigms. *Ecological Economics* **3**, 193–213.

Colchester, M. 1993. Pirates, squatters and poachers: the political ecology of dispossession of the native peoples of Sarawak. *Global Ecology and Biogeography Letters* **3**(4–6), 158–79.

Cooke, R. V. & J. C. Doornkamp 1993. *Geomorphology in environmental management*.

Oxford: Oxford University Press.

Corbridge, S. 1986. *Capitalist world development – a critique of radical development geography*. London: Macmillan.

Cosgrove, D. 1994. Contested global visions: one-world, whole-earth, and the Apollo space photographs. *Annals of the Association of American Geographers* **84**, 270–94.

Cotgrove, S. 1982. *Catastrophe or cornucopia: the environment, politics and society*. Chichester: John Wiley.

Cowell, R. & P. Jehlicka 1995. Backyard and biosphere: the spatial distribution of support for English and Welsh environmental organisations. *Area* **27**(2), 110–17.

Crandall, R. 1980. Motivations for leisure. *Journal of Leisure Research* **12**(1), 45–54.

Cronon, W. 1983. *Changes in the land: Indians, colonists, and the ecology of New England*. New York: Hill & Wang.

Cross, A. M., J. J. Settle, N. A. Drake, R. T. M. Paivinen 1991. Subpixel measurement of tropical forest cover using AVHRR data. *International Journal of Remote Sensing* **12**(5), 1119–29.

Cumberland, K. B. 1941. A century's change: natural to cultural vegetation in New Zealand. *Geographical Review* **31**(4), 525–54.

Cumberland, K. B. 1961. Man In nature in New Zealand. *New Zealand Geographer* **17**(2), 137–54.

Cunningham, W. P. & B. W. Saigo 1992. *Environmental science: a global concern*. Dubuque, Iowa: Brown.

Dauvergne, P. 1993/4. The politics of deforestation in Indonesia. *Pacific Affairs* **66**, 497–518.

Davis, J. R. & P. M. Nanninga 1985. GEOMYCIN: towards a geographic expert system for resource management. *Journal of Environmental Management* **20**(4), 377–90.

DeLeeuw, P. N. & J. C. Tothill 1990. *The concept of rangeland carrying capacity in Subsaharan Africa – myth or reality?* Paper 29b, Pastoral Development Network, Overseas Development Institute, London.

Denslow, J. S. & C. Padoch (eds) 1988. *People of the tropical rainforest*. Berkeley: University of California Press.

DeSanto, R. S. 1976. The Journal's policy and objectives. *Environmental Management* **1**(1), 3.

Devall, W. 1988. *Simple in means, rich in ends: practising deep ecology*. Salt Lake City: Peregrine.

Devall, B. & G. Sessions 1985. *Deep ecology: living as if nature mattered*. Salt Lake City: Peregrine Smith.

Dieter, H. H. 1992. German drinking water regulations, pesticides, and axiom of concern. *Environmental Management* **16**(1), 21–32.

Dobson, A. 1995. *Green political thought*, 2nd edn. London: Routledge.

Doel, M. A. & D. B. Clarke 1996. A night at the movies: the virtual, psychoanalysis, and everyday life. Paper presented to the Social and Cultural Study Group session "Virtual geographies", Annual Conference of the Royal Geographical Society and the Institute of British Geographers, University of Strathclyde, January 1996.

Dorney, R. S. 1987. *The professional practice of environmental management*. New York: Springer.

Dove, M. R. 1983. The agroecological mythology of the Javanese and the political economy of Indonesia. *Indonesia* **39**, 1–36.

Dovers, S. R. & J. W. Handmer 1993. Contradictions in sustainability. *Environmental Conservation* **20**(3), 217–22.

Dunlap, R. E. 1993. From environmental to ecological problems. In *Social problems*, C. Calhoun & G. Ritzer (eds), 707–738. New York: McGraw-Hill.

Durning, A. T. 1993. Supporting indigenous peoples. In *State of the world 1993*, L. R. Brown

(ed.), 80–100. London: Earthscan.
DuVair, P. & J. Loomis 1993. Household's valuation of alternative levels of hazardous waste risk reductions: an application of the referendum format contingent valuation method. *Journal of Environmental Management* **39**(2), 143–52.
Dwyer, L. 1986. Environmental policy and the economic value of human life. *Journal of Environmental Management* **22**(3), 229–44.

Eagly, A. & S. Chaiken 1992. *The psychology of attitudes*. San Diego: Harcourt Brace Jovanovich.
Eccleston, B. 1996. Does North–South collaboration enhance NGO influence on deforestation policies in Malaysia and Indonesia? *Journal of Commonwealth and Comparative Politics* **34**(1), 17–34.
Eckersley, R. 1992. *Environmentalism and political theory: toward an ecocentric approach*. Sydney: Allen & Unwin.
—1993. Free market environmentalism: friend or foe? *Environmental Politics* **2**(1), 1–19.
—(ed.) 1996a. *Markets, the state and the environment – towards integration*. London: Macmillan.
—1996b. Markets, the state and the environment: an overview. In *Markets, the state and the environment – towards integration*, R. Eckersley (ed.), 7–45. London: Macmillan.
Ecologist, The 1972. Blueprint for survival. *The Ecologist* **2**(1), 1–43.
—1993. *Whose common future? Reclaiming the commons*. London: Earthscan.
Eden, S. E. 1993. Individual environmental responsibility and its role in public environmentalism. *Environment and Planning A* **25**, 1743–58.
—1994. Using sustainable development: the business case. *Global Environmental Change* **4**, 160–67.
Edge, G. & K. Tovey 1995. Energy: hard choices ahead. In *Environmental science for environmental management*, T. O'Riordan (ed.), 317–34. Harlow: Longman.
Edwards, K. J. 1988. The hunter–gatherer/agricultural transition and the pollen record in the British Isles. In *The cultural landscape – past, present and future*, H. H. Birks, H. J. B. Birks, P. E. Kaland, D. Moe (eds), 255–66. Cambridge: Cambridge University Press.
Ehrlich, P. R. 1970. *The population bomb*. New York: Ballantine.
Ehrlich, P. R. & A. H. Ehrlich 1990. *The population explosion*. New York: Doubleday.
Ekins, P. E. 1993. Making development sustainable. In *Global ecology*, W. Sachs (ed.), 91–103. London: Zed Books.
El Serafy, S. & E. Lutz 1989. Environmental and natural resource accounting. In *Environmental management and economic development*, G. Schramm & J. Warford (eds), 23–38. Baltimore: Johns Hopkins University Press.
Ennos, A. R. & S. E. R. Bailey 1995. *Problem solving in environmental biology*. London: Longman.
Enzensberger, H. M. 1974. A critique of political ecology. *The New Left Review* **84**, 3–31.
Esteva, G. & M. S. Prakash 1992. Grass-roots resistance to sustainable development. *The Ecologist* **22**, 45–51.
European Commission 1992. Council Regulation no. 2078/92/EEC on agricultural methods compatible with the requirements of the protection of the environment and the maintenance of the countryside. *Official Journal of the European Commission* **L215**, 85–90.

Fairhead, J. & M. Leach 1996. *Reversing landscape history – power, policy and socialised ecology in West Africa's forest–savanna mosaic*. Chichester: John Wiley.
Fairlie, S. (ed.) 1995. Overfishing: its causes and consequences. *The Ecologist* [special double issue], **25**(2/3), 42–125.
Fisher, J. 1993. *The road from Rio: sustainable development and the non-governmental movement*

in the Third World. Westport, Connecticut: Praeger.
Flannery, T. F. 1990. Pleistocene faunal loss: implications of the aftershock for Australia's past and future. *Archaeology in Oceania* **25**, 45–67.
Fleet, H. 1984. *New Zealand's forests*. Auckland: Heinemann.
Flynn, A. & P. Lowe 1992. The greening of the Tories: the Conservative Party and the environment. In *Green politics two*, W. Ruedig (ed.), 9–36. Edinburgh: Edinburgh University Press.
FOE 1992. *The rainforest harvest*. London: Friends of the Earth.
—1994a. *Planning for the planet*. London: Friends of the Earth.
—1994b. *Mahogany is murder* [report L232]. London: Friends of the Earth.
Foran, B. D. & D. M. Stafford Smith 1991. Risk, biology and drought management strategies for cattle stations in central Australia. *Journal of Environmental Management* **33**(1), 17–34.
Foreman, D. & B. Haywood (eds) 1988. *Ecodefense: a field guide to monkeywrenching*, 2nd edn. Tucson: Ned Ludd Books.
Franke, R. W. & B. H. Chasin 1980. *Seeds of famine: ecological destruction and the development dilemma in the West African Sahel*. Montclair, New Jersey: Allanheld & Osman.
Frawley, K. J. 1987. *The Malaan group settlement North Queensland 1954*. Monograph Series 2, Department of Geography and Oceanography, Australian Defence Force Academy, Canberra.
Freeman, D. M. & R. S. Frey 1986. Method for assessing the social impacts of natural resource policies. *Journal of Environmental Management* **23**(3), 229–46.
Furnivall, J. S. 1956. *Colonial policy and practice: a comparative study of Burma and Netherlands India*. New York: New York University Press.

Gadgil, M., F. Berkes, C. Folke 1993. Indigenous knowledge for biodiversity conservation. *Ambio* **22**(2/3), 151–6.
Garlauskas, A. B. 1975. Conceptual framework of environmental management. *Journal of Environmental Management* **3**(3), 185–203.
Garner, R. 1996. *Environmental politics*. London: Harvester–Wheatsheaf.
Garrido, F. & E. Moyano 1996. Spain. In *The European environment and CAP reform: policies and prospects for conservation*, M. Whitby (ed.), 86–104. Wallingford: CAB International.
George, S. & F. Sabelli 1994. *Faith and credit: the World Bank's secular empire*. Harmondsworth: Penguin.
Gerrard, S. 1995. Environmental risk management. In *Environmental science for environmental management*, T. O'Riordan (ed.), 296–316. Harlow: Longman.
Gibson, W. 1988. *Burning chrome*. London: Grafton.
—1993. *Neuromancer*. London: HarperCollins.
Giddens, A. 1979. *Central problems in social theory: action, structure and contradiction in social analysis*. London: Macmillan.
Gillespie, R., D. R. Horton, P. Ladd, T. H. Rich, R. Thorne, R. V. S. Wright 1978. Lancefield Swamp and the extinction of the Australian megafauna. *Science* **200**, 1044–1048.
Glacken, C. 1967. *Traces on the Rhodian Shore*. Berkeley: University of California Press.
Gleick, J. 1987. *Chaos – the making of a new science*. New York: Viking.
Goodman, D. & M. R. Redclift 1991. *Refashioning nature: food, ecology and culture*. London: Routledge.
Graham, S. & S. Marvin 1996. *Telecommunications and the city: electronic spaces, urban places*. London: Routledge.
Grainger, A. 1993. *Controlling tropical deforestation*. London: Earthscan.
Grant, A. & T. Jickells 1995. Marine and estuarine pollution. In *Environmental science for environmental management*, T. O'Riordan (ed.), 263–82. Harlow: Longman.
Gray, D. B. 1985. *Ecological beliefs and behaviours*. London: Greenwood.

Grove, R. 1993. Conserving Eden: the (European) East India Companies and their environmental policies on St Helena, Mauritius and in Western India, 1660 to 1854. *Comparative Studies in Society and History* **35**, 318–51.

Grubb, M., M. Koch, A. Munson, F. Sullivan, K. Thomson (eds) 1993. *The Earth Summit agreements: a guide and assessment*. London: Earthscan.

Guha, R. 1989. *The unquiet woods: ecological change and peasant resistance in the Himalaya*. Delhi: Oxford University Press.

Gumbricht, T. 1996. Application of GIS in training for environmental management. *Journal of Environmental Management* **46**(1), 17–30.

Haas, P. M. 1990. *Saving the Mediterranean: the politics of international environmental cooperation*. New York: Columbia University Press.

Hall, J. A. (ed.) 1986. *States in history*. Oxford: Basil Blackwell.

Hall, C. M. 1988. The "worthless lands hypothesis" and Australia's national parks and reserves. In *Australia's ever changing forests*, K. J. Frawley & N. Semple (eds), 441–58. Canberra: Australian Defence Force Academy Press.

Hammit, W. 1993. Use patterns and solitude preferences of shelter campers in Great Smokey Mountains National Park. *Journal of Environmental Management* **38**, 45–53.

Hardin, G. 1968. The tragedy of the commons. *Science* **162**, 1243–8.

Hardoy, J. E., D. Mitlin, D. Satterwaite 1992. *Environmental problems in Third World cities*. London: Earthscan.

Harris, D. R. 1978. The environmental impact of traditional and modern agricultural systems. In *Conservation and agriculture*, J. G. Hawkes (ed.), 61–9. Montclair, New Jersey: Allanheld.

Harrison, P. 1993. *The third revolution: population, environment and a sustainable world*. Harmondsworth: Penguin.

Harvey, D. 1974. Population, resources and the ideology of science. *Economic Geography* **50**, 256–77.

—1989. *The conditions of postmodernity: an enquiry into the origins of cultural change*. Oxford: Basil Blackwell.

—1993. The nature of environment: the dialectics of social and environmental change. In *Real problems – false solutions: socialist register 1993*, R. Miliband & L. Panitch (eds), 1–51. London: Merlin Press.

Hawkes, J. G. (ed.) 1978. *Conservation and agriculture*. Montclair, New Jersey: Allanheld.

Hays, S. P. 1987. *Beauty, health and permanence: environmental politics in the United States, 1955–1985*. Cambridge: Cambridge University Press.

Head, L. 1989. Prehistoric Aboriginal impacts on Australian vegetation: an assessment of the evidence. *Australian Geographer* **20**, 37–46.

Head, L. 1993. Unearthing prehistoric cultural landscapes: a view from Australia. *Transactions of the Institute of British Geographers* **18**(4), 481–99.

Headrick, D. R. 1981. *The tools of empire: technology and European imperialism in the nineteenth century*. Oxford: Oxford University Press.

Hecht, S. B. & A. Cockburn 1989. *The fate of the forests: developers, destroyers and defenders of the Amazon*. London: Verso.

Heilbroner, R. 1974. *An inquiry into the human prospect*. New York: Norton.

Heske, F. 1938. *German forestry*. New Haven, Connecticut: Yale University Press.

Hill, S. 1988. *The tragedy of technology*. London: Pluto Press.

Hilton, M. J. 1994. Applying the principle of sustainability to coastal sand mining: the case of Pakiri–Mangawhai Beach, New Zealand. *Environmental Management* **18**(6), 815–30.

Hoggart, K., H. Buller, R. Black 1995. *Rural Europe – identity and change*. London: Edward Arnold.

Hollick, M. 1981. The role of quantitative decision-making methods in environmental impact assessment. *Journal of Environmental Management* **12**(1), 65–78.
Holloway, N. 1995. No pain, no grain. *Far Eastern Economic Review* **158**(46), 88–91.
Homewood, K. & W. A. Rodgers 1987. Pastoralism, conservation and the overgrazing controversy. In *Conservation in Africa: people, policies and practice*, D. Anderson & R. Grove (eds), 111–28. Cambridge: Cambridge University Press.
Hong, E. 1987. *Natives of Sarawak: survival in Borneo's vanishing forests*. Penang: Institut Masyarakat.
Hooper, P. 1981. *Our forests – ourselves*. Dunedin, New Zealand: McIndoe.
Horowitz, M. M. & P. D. Little 1987. African pastoralism and poverty: some implications for drought and famine. In *Drought and hunger in Africa: denying famine a future*, M. H. Glantz (ed.), 59–82. Cambridge: Cambridge University Press.
Horton, D. R. 1982. The burning question: Aborigines, fire and Australian ecosystems. *Mankind* **13**, 237–51.
Hoskins, W. G. 1955. *The making of the English landscape*. Harmondsworth: Penguin.
Hurrell, A. 1992. Brazil and the international politics of Amazonian deforestation. In *The international politics of the environment*, A. Hurrell & B. Kingsbury (eds), 398–429. Oxford: Oxford University Press.
Hurrell, A. 1994. A crisis of ecological viability? Global environmental change and the nation state. *Political Studies* **42**, 146–65.
Hurrell, A. & B. Kingsbury (eds) 1992. *The international politics of the environment: actors, interests, and institutions*. Oxford: Oxford University Press.
Hurst, P. 1990. *Rainforest politics: ecological destruction in South East Asia*. London: Zed Books.
Hutchinson, C. 1995. *Vitality and renewal: a manager's guide for the 21st century*. New York: Praeger.

ICC 1990. *The business charter for sustainable development: principles for environmental management*. Paris: International Chamber of Commerce.
Inglehart, R. 1977. *The silent revolution: changing values and political style among Western publics*. Princeton, New Jersey: Princeton University Press.
Ingold, T., D. Riches, J. Woodburn (eds) 1988. *Hunters and gatherers*, vol. I: *history, evolution and social change*. Oxford: Berg.

Jackson, C. 1994. Gender analysis and environmentalism. In *Social theory and the global environment*, M. Redclift & T. Benton (eds), 113–49. London: Routledge.
Jacob, M. 1994. Sustainable development and deep ecology: an analysis of competing traditions. *Environmental Management* **18**(4), 477–88.
Jahn, T. 1993. Ecological movements and environmental politics in Germany. *Capitalism Nature Socialism* **4**(1), 1–10.
James, N. D. G. 1981. *A history of English forestry*. Oxford: Basil Blackwell.
Jeffers, J. N. R. 1973. Editorial introduction. *Journal of Environmental Management* **1**, 1–2.
Jewitt, S. 1995. Europe's "others"? Forestry policy and practices in colonial and post-colonial India. *Environment and Planning D* **13**, 67–90.
Jickells, T. D., R. Carpenter, P. S. Liss 1990. Marine environment. In *The Earth as transformed by human action – global and regional changes in the biosphere over the past 300 years*, B. L. Turner, W. C. Clark, R. W. Kates, J. F. Richards, J. T. Mathews, W. B. Meyer (eds) 313–34. Cambridge: Cambridge University Press.
Johnson, S. 1994. Turning the tide. *Geographical Magazine* **66**(9), 12–14.
Johnston, R. J. 1983. Resource analysis, resource management and the integration of physical and human geography. *Progress in Physical Geography* **7**, 127–46.
—1989. *Environmental problems: nature, society, economy*. London: Pinter (Belhaven).

—1992. Laws, states and super-states: international law and the environment. *Applied Geography* **12**, 211–28.

Jones, A. 1994. *The new Germany*. Chichester: John Wiley.

Jones, R. 1991. Landscapes of the mind: Aboriginal perceptions of the natural world. In *The humanities and the Australian environment*, D. J. Mulvaney (ed.), 21–48. Canberra: Australian Academy of Humanities.

Jordan, A. 1993. Integrated pollution control and the evolving style and structure of environmental regulation in the UK. *Environmental Politics* **2**(3), 405–427.

Judelson, M. 1986. The proliferation of the chain saw and its implications on forest lands. *Journal of Environmental Management* **22**(2), 95–104.

Karshenas, M. 1994. Environment, technology and employment: towards a new definition of sustainable development. *Development and Change* **25**, 723–56.

Kellert, S. R. 1984. Urban American perceptions of animals and the natural environment. *Urban Ecology* **8**, 209–228.

Keohane, R. O. & E. Ostrom (eds) 1994. *Local commons and global interdependence*. London: Sage.

Kerrod, R. 1990. *The journeys of Voyager – NASA reaches for the planets*. London: Prion.

Kershaw, A. P. 1986. Climatic change and Aboriginal burning in north-east Australia during the last two glacial/interglacial cycles. *Nature* **322**, 47–9.

Khasiani, S. A. (ed.) 1992. *Groundwork: African women as environmental managers*. Nairobi: African Centre for Technology Studies.

Kliskey, A. D. 1994. A comparative analysis of approaches to wilderness perception mapping. *Journal of Environmental Management* **41**, 199–236.

Kloppenburg, J. R. 1988. *First the seed: the political economy of plant biotechnology, 1492–2000*. Cambridge: Cambridge University Press.

Kneale, J. 1996. Conceiving the inconceivable: reading Gibson's cyberspace. Paper presented to the Social and Cultural Study Group session "Virtual geographies", Annual Conference of the Royal Geographical Society and the Institute of British Geographers, University of Strathclyde, January 1996.

Knickel, K. 1990. Agricultural structural change: impact on the rural environment. *Journal of Rural Studies* **6**(4), 383–93.

Korten, D. C. 1995. *When corporations rule the world*. London: Earthscan.

Kreutzwiser, R. D. & L. J. Pietraszko 1986. Wetland values and protection strategies: a study of landowner attitudes in Southern Ontario. *Journal of Environmental Management* **22**, 13–23.

Krol, A. 1995. Environmental management – issues and approaches for an organization. In *Business and the environment*, M. D. Rogers (ed.), 51–88. London: Macmillan.

Kuhn, T. S. 1970. *The structure of scientific revolutions*, 2nd edn. Chicago: University of Chicago Press.

Kuitunen, M. & T. Tormala 1994. Willingness of students to favour the protection of endangered species in a trade-off conflict in Finland. *Journal of Environmental Management* **42**(2), 111–18.

Kummer, P. M. 1992. *Deforestation in the postwar Philippines*. Manila: Ateneo de Manila University Press.

Laferriere, E. 1994. Environmentalism and the global divide. *Environmental Politics* **3**(1), 91–113.

Lash, S., B. Szerszynski, B. Wynne (eds) 1996. *Risk, environment and modernity: towards a new ecology*. London: Sage.

LeHeron, R. 1988. Food and fibre production under capitalism – a conceptual agenda. *Progress in Human Geography* **12**(3), 409–430.

Lele, S. M. 1991. Sustainable development: a critical review. *World Development* **19**, 607–621.

Leonard, H. J. 1988. *Pollution and the struggle for the world product: multinational corporations, environment and international comparative advantage*. Cambridge: Cambridge University Press.

LePrestre, P. G. 1989. *The World Bank and the environmental challenge*. London: Associated University Presses.

Lester, J. P. (ed.) 1990. *Environmental politics and policy: theories and evidence*. Durham, North Carolina: Duke University Press.

Light, J. S. 1996. Developing the virtual landscape. *Environment and Planning D* **14**, 127–31.

Lipschutz, R. D. & K. Conca (eds) 1993. *The state and social power in global environmental politics*. New York: Columbia University Press.

Little, P. D. 1987. Land use conflicts in the agricultural/pastoral borderlands: the case of Kenya. In *Lands at risk in the Third World*, P. D. Little & M. M. Horowitz (eds), 195–212. Boulder, Colorado: Westview.

Little, P. D. & M. Watts (eds) 1994. *Living under contract: contract farming and agrarian transformation in sub-Saharan Africa*. Madison: University of Wisconsin Press.

Lohmann, L. 1990. Commercial tree plantations in Thailand: deforestation by any other name. *The Ecologist* **20**(1), 9–17.

—1996. Freedom to plant – Indonesia and Thailand in a globalizing pulp and paper industry. In *Environmental change in South-East Asia: people, politics and sustainable development*, M. Parnwell & R. L. Bryant (eds), 23–48. London: Routledge.

Lorrain-Smith, R. 1982. The nature of environmental management. *Journal of Environmental Management* **14**(3), 229–36.

Lovelock, J. E. 1995. *Gaia: a new look at life on Earth*, 3rd edn. New York: Oxford University Press.

Lowe, M. S. & S. R. Bowlby 1992. Population and environment. In *Environmental issues in the 1990s*, A. M. Mannion & S. R. Bowlby (eds), 117–30. Chichester: John Wiley.

Lowe, P. & J. Goyder 1983. *Environmental groups in politics*. London: Allen & Unwin.

Lowe, P., G. Cox, M. MacEwen, T. O'Riordan, M. Winter 1986. *Countryside conflicts: the politics of farming, forestry and conservation*. Aldershot: Gower.

Luke, T. W. 1995. Reproducing Planet Earth? The hubris of Biosphere 2. *The Ecologist* **25**, 157–62.

Lutz, W. (ed.) 1994. *The future population of the world*. London: Earthscan.

Lyon, V. 1992. Green politics: political parties, elections and environmental policy. In *Canadian environmental policy: ecosystems, politics and process*, R. Boardman (ed.), 126–43. Toronto: Oxford University Press.

Mack, P. 1990. *Viewing the Earth: the social construction of the Landsat satellite system*. Cambridge, Massachusetts: MIT Press.

MacNeill, J. W. 1971. *Environmental management*. Ottawa: Information Canada.

Mann, M. 1986. *The sources of social power, I – a history of power from the beginning to AD 1760*. Cambridge: Cambridge University Press.

Mannion, A. M. 1992. Acidification and eutrophication. In *Environmental issues in the 1990s*, A. M. Mannion & S. R. Bowlby (eds) 177–96. Chichester: John Wiley.

Marchak, M. P. 1995. *Logging the globe*. Montreal : McGill–Queen's University Press.

Marsden, T., J. Murdoch, P. Lowe, R. Munton, A. Flynn 1993. *Constructing the countryside*. London: UCL Press.

Marsh, G. 1965 (1864). *Man and nature; or physical geography as modified by human action*. Cambridge, Massachusetts: Harvard University Press.

Martin, P. S. & R. G. Kline (eds) 1984. *Quaternary extinctions: a prehistoric revolution*. Tucson: University of Arizona Press.

Marx, K. 1973 (1857–8). *Grundrisse*. Harmondsworth: Penguin.
Maser, C. 1994. *Sustainable forestry: philosophy, science and economics*. New York: St Lucie Press.
Mather, A. S. & K. Chapman 1995. *Environmental resources*. London: Longman.
Mazurski, K. R. 1990. Industrial pollution: the threat to Polish forests. *Ambio* **19**, 70–74.
McAllister, I. 1994. Dimensions of environmentalism: public opinion, political activism, and party support in Australia. *Environmental Politics* **3**(1), 22–42.
McCormick, J. 1995. *The global environmental movement*, 2nd edn. Chichester: John Wiley.
McDowell, C. & R. Sparks 1989. The multivariate modelling and prediction of farmers' conservation behaviour towards natural ecosystems. *Journal of Environmental Management* **28**, 185–210.
McDowell, M. A. 1989. Development and the environment in ASEAN. *Pacific Affairs* **62**, 307–329.
McEachern, C. 1992. Farmers and conservation: conflict and accommodation in farming politics. *Journal of Rural Studies* **8**(2), 159–71.
McGlone, M. S. 1983. Polynesian deforestation of New Zealand – a preliminary synthesis. *Archaeology in Oceania* **18**, 11–25.
McGregor, W. R. 1948. *The Waipoua kauri forest of northern New Zealand*. Auckland: Billingstone.
Meadows, D. L., J. Randers, W. W. Behrens 1972. *The limits to growth: a report to the Club of Rome's project on the predicament of mankind*. New York: Potomac Associates.
Meggars, B. J. 1971. *Amazonia: man and culture in a counterfeit paradise*. Chicago: Aldine.
Memon P. A. & G. A. Wilson 1993. Indigenous forests. In *Environmental planning in New Zealand*, P. A. Memon & H. C. Perkins (eds), 97–119. Palmerston North: Dunmore.
Merchant, C. 1982. *The death of nature: women, ecology and the scientific revolution*. London: Wildwood House.
Micklewright, S. 1993. The voluntary movement. In *Conservation in progress*, F. B. Goldsmith & A. Warren (eds), 321–34. Chichester: John Wiley.
Middleton, N., P. O'Keefe, S. Mayo 1993. *The tears of the crocodile: from Rio to reality in the developing world*. London: Pluto Press.
Miller, A. 1984. Professional collaboration in environmental management: the effectiveness of expert groups. *Journal of Environmental Management* **18**(4), 365–88.
—1985. Technological thinking: its impact on environmental management. *Environmental Management* **9**(3), 179–90.
—1993. The role of analytical science in natural resource decision making. *Environmental Management* **17**(5), 563–74.
Miller, G. T. 1994. *Living in the environment: principles, connections, and solutions*, 8th edn. Belmont, California: Wadsworth.
Miller, M. A. L. 1995. *The Third World in global environmental politics*. Boulder, Colorado: Lynne Rienner.
Mische, P. M. 1989. Ecological security and the need to reconceptualize sovereignty. *Alternatives* **14**, 389–427.
Mitchell, B. 1989. *Geography and resource analysis*, 2nd edn. New York: Longman.
—1995. *Resource and environmental management in Canada: addressing conflict and uncertainty*, 2nd edn. Oxford: Oxford University Press.
Moffat, I. 1990. The potentialities and problems associated with applying information technology to environmental management. *Journal of Environmental Management* **30**(3), 209–220.
—1992. The evolution of the sustainable development concept: a perspective from Australia. *Australian Geographical Studies* **30**(1), 27–42.
Mohnen, V. 1988. The challenge of acid rain. *Scientific American* **159**(2), 30–38.
Moody, R. 1996. Mining the world: the global reach of Rio Tinto Zinc. *The Ecologist* **26**(2),

46–52.
Morehouse, W. 1994. Unfinished business: Bhopal ten years after. *The Ecologist* **24**, 164–8.
Morris, C. & C. Potter 1995. Recruiting the new conservationists: farmers adoption of agri-environmental schemes in the UK. *Journal of Rural Studies* **11**(1), 51–63.
Mould, R. F. 1992. *Chernobyl: the real story*. Oxford: Pergamon Press.
Murdoch, J. & T. Marsden 1995. The spatialization of politics: local and national actor-spaces in environmental conflict. *Transactions of the Institute of British Geographers* **20**, 368–80.
Mwaloyosi, R. P. B. 1991. Population growth, carrying capacity and sustainable development in south-west Masailand. *Journal of Environmental Management* **33**(2), 175–88.
Myers, N. & J. Simon 1994. *Scarcity or abundance? A debate on the environment*. New York: Norton.
Myerson, G. & Y. Rydin 1996. *The language of environment: a new rhetoric*. London: UCL Press.

Naess, A. 1989. *Ecology, community and life-style: outline of an ecosophy*. Cambridge: Cambridge University Press.
Nash, R. 1967. *Wilderness and the American mind*. New Haven, Connecticut: Yale University Press.
Nath, B., L. Hens, P. Compton, D. Devuyst 1993. *Environmental management* [3 vols]. Brussels: VUB Press.
Nemecek, S. 1996. Virtual pollution. *Scientific American* **274**, 13–14.
North, R. 1995. *Life on a modern planet – rediscovering faith in progress*. Manchester: Manchester University Press.

O'Connor, J. 1988. Capitalism, nature, socialism: a theoretical introduction. *Capitalism Nature Socialism* **1**, 11–38.
O'Connor, M. 1994. Introduction: liberate, accumulate – and bust? In *Is capitalism sustainable: political economy and the politics of ecology*, M. O'Connor (ed.), 1–21. London: Guilford Press.
O'Connor, R. & M. Shrubb 1988. *Farming and birds*. Cambridge: Cambridge University Press.
Odell, P. R. 1983. *Oil and world power*, 2nd edn. Penguin: Harmondsworth.
Odum, E. P. 1971. *Fundamentals of ecology*. Philadelphia: W. B. Saunders.
Olofson, H. 1995. Taboo and environment, Cebuano and Tagbanuwa: two cases of indigenous management of natural resources in the Philippines and their relation to religion. *Philippine Quarterly of Culture and Society* **23**, 20–34.
O'Neill, J. 1993. *Ecology, policy and politics: human wellbeing and the natural world*. London: Routledge.
O'Riordan, T. 1971. *Perspectives on resource management*. London: Pion.
—1981. *Environmentalism*. London: Pion.
—(ed.) 1995a. *Environmental science for environmental management*. Harlow: Longman.
—1995b. Frameworks for choice: core beliefs and the environment. *Environment* **37**(8), 4–9, 25–29.
O'Riordan, T. & J. Cameron (eds) 1994. *Interpreting the precautionary principle*. London: Earthscan.
Ostrom, E. 1990. *Governing the commons: the evolution of institutions for collective action*. Cambridge: Cambridge University Press.

Park, C. C. 1980. *Ecology and environmental management*. Boulder, Colorado: Westview.
Parnwell, M. J. G. & V. T. King 1995. Environmental degradation, resource scarcity and population movement among the Iban of Sarawak. Paper presented at the conference

of the European Association of South East Asian Studies, June–July 1995, Leiden, Netherlands.

Passmore, J. 1980. *Man's responsibility to nature*, 2nd edn. London: Duckworth.

Parry, M. 1990. *Climate change and world agriculture*. London: Earthscan.

Pearce, D., A. Marhandya, E. Barbier 1989. *Blueprint for a green economy*. London: Earthscan.

Pearson, C. S. (ed.) 1987. *Multinational corporations, environment, and the Third World*. Durham, North Carolina: Duke University Press.

Peet, R. 1991. *Global capitalism: theories of societal development*. London: Routledge.

Peet, R. & M. Watts 1993. Introduction: development theory and environment in an age of market triumphalism. *Economic Geography* **69**(3), 227–53.

Peluso, N. L. 1992. *Rich forests, poor people: resource control and resistance in Java*. Berkeley: University of California Press.

—1993. Coercing conservation: the politics of state resource control. In *The state and social power in global environmental politics*, R. D. Lipschutz & K. Conca (eds), 46–70. New York: Columbia University Press.

—1995. Whose woods are these? Counter-mapping forest territories in Kalimantan, Indonesia. *Antipode* **27**(4), 383–406.

Pepper, D. 1984. *The roots of modern environmentalism*. London: Croom Helm.

—1993. *Ecosocialism: from deep ecology to social justice*. London: Routledge.

PER 1992. *The future of people and forests in Thailand after the logging ban*. Bangkok: Project for Ecological Recovery.

Perry, R. W. & L. Nelson 1991. Ethnicity and hazard information dissemination. *Environmental Management* **15**(4), 581–8.

Petak, W. J. 1980. Environmental planning and management: the need for an integrative perspective. *Environmental Management* **4**(4), 287–96.

—1981. Environmental management: a system approach. *Environmental Management* **5**(3), 213–24.

Pickering, K. T. & L. A. Owen 1994. *An introduction to global environmental issues*. London: Routledge.

Pitman, J. I. 1992. Changes in crop productivity and water quality in the United Kingdom. In *Agricultural change, environment and economy*, K. Hoggart (ed.), 89–122. London: Cassell (Mansell).

Poffenberger, M. (ed.) 1990. *Keepers of the forest: land management alternatives in Southeast Asia*. West Hartford, Connecticut: Kumerian Press.

Polanyi, K. 1957 (1944). *The great transformation: the political and economic origins of our time*. Boston: Beacon Press.

Porritt, J. 1990. *Friends of the Earth handbook*. London: Macdonald.

Porter, G. & J. W. Brown 1991. *Global environmental politics*. Boulder, Colorado: Westview.

Potter, C. 1990. Conservation under a European farm survival policy. *Journal of Rural Studies* **6**(1), 1–7.

Potter, C., H. Cook, C. Norman 1993. The targeting of rural environmental policies: an assessment of agri-environmental schemes in the UK. *Journal of Environmental Planning and Management* **36**, 199–216.

Powell, J. M. 1976. *Environmental management in Australia 1788–1914*. Melbourne: Oxford University Press.

Pretty, J. N. 1995. *Regenerating agriculture – policies and practice for sustainability and self-reliance*. London: Earthscan.

Prigogine, I. & I. Stangers 1985. *Order out of chaos: man's new dialogue with nature*. London: Flamingo.

Princen, T. & M. Finger 1994. *Environmental NGOs in world politics: linking the local and the global*. London: Routledge.

Putz, F. E. & N. M. Holbrook 1988. Tropical rainforest images. In *People of the tropical rainforest*, J. S. Denslow & C. Padoch (eds), 37–52. Berkeley: University of California Press.

Reckow, K. H. 1994. Importance of scientific uncertainty in decision making. *Environmental Management* **18**(2), 161–6.

Redclift, M. 1987. *Sustainable development: exploring the contradictions*. London: Methuen.

Redclift, M. & G. Woodgate 1994. Sociology and the environment: discordant discourse? In *Social theory and the global environment*, M. Redclift & T. Benton (eds), 51–66. London: Routledge.

Rees, J. 1990. *Natural resources: allocation, economics and policy*, 2nd edn. London: Routledge.

Reichel-Dolmatoff, G. 1996. *The forest within: the world-view of the Tukano Amazonian Indians*. London: Themis.

Reid, D. 1995. *Sustainable development*. London: Earthscan.

Rettig, R. B. 1995. Management regimes in ocean fisheries. In *The handbook of environmental economics*, D. W. Bromley (ed.), 433–52. Oxford: Blackwell.

Ribot, J. C. 1993. Market–state relations and environmental policy – limits of state capacity in Senegal. In *The state and social power in global environmental politics*, R. D. Lipschutz & K. Conca (eds), 24–45. New York: Columbia University Press.

Rich, B. 1994. *Mortgaging the earth: the World Bank, environmental impoverishment and the crisis of development*. London: Earthscan.

Rigg, J. 1991. *Southeast Asia – a region in transition*. London: Unwin Hyman.

Robertson, W. A. 1993. New Zealand's new legislation for sustainable resource management: the Resource Management Act 1991. *Land Use Policy* **10**(4), 303–311.

Robinson, G. M. 1991. EC agricultural policy and the environment – land use implications in the UK. *Land Use Policy* **8**(2), 95–107.

Robinson, G. M. & B. W. Ilbery 1993. Reforming the CAP: beyond MacSharry. *Progress in Rural Policy and Planning* **3**, 197–207.

Robinson, M. 1992. *The greening of British party politics*. Manchester: Manchester UniversityPress.

Roche, M. M. 1990. *History of New Zealand forestry*. Palmerston North: GP Publications.

Rowell, A. 1995. Oil, Shell and Nigeria. *The Ecologist* **25**(6), 210–13.

Rowlands, I. 1995. *The politics of global atmospheric change*. Manchester: Manchester University Press.

Rozanov, B. G., V. Targulian, D. S. Orlov 1990. Soils. In *The Earth as transformed by human action – global and regional changes in the biosphere over the past 300 years*, B. L. Turner, W. C. Clark, R. W. Kates, J. F. Richards, J. T. Mathews, W. B. Meyer (eds), 203–214. Cambridge: Cambridge University Press.

RTZ [Rio Tinto Zinc] 1994. The environment in focus. *RTZ Review* **31**, 3–5.

Rush, J. 1991. *The last tree – reclaiming the environment in tropical Asia*. New York: Asia Society.

Sachs, W. (ed.) 1993. *Global ecology: a new area of political conflict*. London: Zed Books.

Sandhu, H. S. 1977. A definition of environmental research. *Environmental Management* **1**(6), 483.

Sargent, C. & S. Bass (eds) 1992. *Plantation politics: forest plantations in development*. London: Earthscan.

Sarkar, A. U. & W. McKillop 1994. Economic parameters for resource policy analysis in developing countries. *Journal of Environmental Management* **40**(4), 309–316.

Sarre, P. & J. Blunden (eds) 1995. *An overcrowded world? Population, resources and the environment*. Oxford: Oxford University Press.

Sauer, C. 1956. The agency of man on Earth. In *Man's role in changing the face of the Earth*, W. Thomas (ed.), 49–69. Chicago: University of Chicago Press.

Schafer, E. L. & J. B. Davis 1989. Making decisions about environmental management when conventional economic analysis cannot be used. *Environmental Management* **13**(2), 189–98.

Schmidheiny, S. 1992. *Changing course: a global perspective on development and the environment*. Cambridge, Massachusetts: MIT Press.

Schramm, G. & J. Warford (eds) 1989. *Environmental management and economic development*. Washington DC: World Bank.

Schumacher, E. F. 1973. *Small is beautiful: a study of economics as if people mattered*. London: Abacus.

Scoones, I. 1989. *Patch use by cattle in dryland Zimbabwe: farmer knowledge and ecological theory*. Paper 28b, Pastoral Development Network, Overseas Development Institute, London.

Scott, J. C. 1976. *The moral economy of the peasant*. New Haven, Connecticut: Yale University Press.

—1985. *Weapons of the weak: everyday forms of peasant resistance*. New Haven, Connecticut: Yale University Press.

Seabrook, J. 1990. *The myth of the market: promises and illusions*. Bideford, Devon: Green Books.

Sen, G. 1987. *Gender and cooperative conflicts*. Helsinki: Wider.

Sessions, G. (ed.) 1994. *Deep ecology for the 21st century*. Boston: Shambhala Press.

Sewell, W. D. R. 1973. Broadening the approach to evaluation in resources management decision-making. *Journal of Environmental Management* **1**(1), 33–60.

Shepard, P. 1969. *English reaction to the New Zealand landscape before 1850*. Wellington: Victoria University Press.

Shiva, V. 1989. *Staying alive: women, ecology and development*. London: Zed Books.

—1991. *Ecology and the politics of survival: conflicts over natural resources in India*. London: Sage.

Short, J. R. 1991. *Imagined country: society, culture and environment*. London: Routledge.

Shultis, J. 1995. Improving the wilderness – common factors in creating national parks and equivalent reserves during the nineteenth century. *Forest and Conservation History* **39**, 121–9.

Silva, E. 1994. Thinking politically about sustainable development in the tropical forests of Latin America. *Development and Change* **25**, 697–721.

Simmons, I. G. 1989. *Changing the face of the Earth: culture, environment, history*. Oxford: Basil Blackwell.

—1993. *Environmental history – a concise introduction*. Oxford: Basil Blackwell.

Simon, J. L. 1981. *The ultimate resource*. Princeton, New Jersey: Princeton University Press.

Simon, J. L. & H. Kahn 1984. *The resourceful Earth*. Cambridge, Massachusetts: Basil Blackwell.

Sklair, L. 1994. Global sociology and global environmental change. In *Social theory and the global environment*, M. Redclift & T. Benton (eds), 205–227. London: Routledge.

Skocpol, T. 1985. Bringing the state back in: strategies of analysis in current research. In *Bringing the state back in*, P. B. Evans, D. Rueschmeyer, T. Skocpol (eds), 3–37. Cambridge: Cambridge University Press.

Skogstad, G. & P. Kopas 1992. Environmental policy in a federal system: Ottawa and the Provinces. In *Canadian environmental policy: ecosystems, politics and process*, R. Boardman (ed.), 43–59. Toronto: Oxford University Press.

Slocombe, D. S. 1993. Environmental planning, ecosystem science, and ecosystem approaches for integrating environment and development. *Environmental Management* **17**(3), 289–304.

Slouka, M. 1995. *War of the worlds: the assault on reality*. London: Abacus.

Stankey, G. H. 1980. Wilderness carrying capacity: management and research progress in the United States. *Landscape Research* **5**(3), 6–11.

Steel, B. S. 1996. Thinking globally and acting locally?: Environmental attitudes, behavior and activism. *Journal of Environmental Management* **47**, 27–36.

Stocking, M. 1987. Measuring land degradation. In *Land degradation and society*, P. Blaikie & H. Brookfield (eds), 49–63. London: Methuen.

Stoddart, D. R. 1981. The paradigm concept and the history of geography. In *Geography, ideology and social concern*, D. R. Stoddart (ed.), 70–80. Oxford: Oxford University Press.

Stoett, P. J. 1993. International politics and the protection of great whales. *Environmental Politics* **2**(2), 277–303.

Swift, A. 1993. *Global political ecology*. London: Pluto Press.

Szerszynski, B., S. Lash, B. Wynne 1996. Introduction: ecology, realism and the social sciences. In *Risk, environment and modernity: towards a new ecology*, S. Lash, B. Szerszynski, B. Wynne (eds), 1–26. London: Sage.

Thapa, G. B. & K. E. Weber 1991. Soil erosion in developing countries: a politico-economic explanation. *Environmental Management* **15**(4), 461–73.

Thomas, K. 1983. *Man and the natural world*. London: Allen Lane.

Thomas, W. L. 1956. *Man's role in changing the face of the Earth*. Chicago: University of Chicago Press.

Tiffen, M., M. Mortimore, F. Gichuki 1994. *More people, less erosion: environmental recovery in Kenya*. Chichester: John Wiley.

Tivers, J. 1996. Landscapes of computer games. Paper presented to the Social and Cultural Study Group session "Virtual geographies", Annual Conference of the Royal Geographical Society and the Institute of British Geographers, University of Strathclyde, January 1996.

Totman, C. 1989. *The green archipelago: forestry in preindustrial Japan*. Berkeley: University of California Press.

Treece, D. 1990. Indigenous peoples in Brazilian Amazonia and the expansion of the economic frontier. In *The future of Amazonia: destruction of sustainable development*, D. Goodman & A. Hall (eds), 264–87. London: Macmillan.

Trudgill, S. 1991. Environmental issues. *Progress in Physical Geography* **15**(1), 84–90.

Tuan, Y. F. 1972. *Topophilia – a study of environmental perceptions, attitudes, and values*. Englewood Cliffs, New Jersey: Prentice-Hall.

Tucker, R. P. & J. F. Richards (eds) 1983. *Global deforestation and the nineteenth century world economy*. Durham, North Carolina: Duke University Press.

Turner, F. J. 1920 (1894). *The frontier thesis in American history*. New York: Holt.

Turner, R. K. (ed.) 1988. *Sustainable environmental management*. Boulder, Colorado: Westview.

—1993. *Sustainable environmental economics and management: principles and practice*, 2nd edn. London: Pinter (Belhaven).

Turner, T. 1989. Kayapo plan meeting to discuss dams. *Cultural Survival Quarterly* **13**, 20–22.

Utting, P. 1993. *Trees, people and power: social dimensions of deforestation and forest protection in Central America*. London: Earthscan.

Van Beek, W. E. A. & P. M. Banga 1992. The dogon and their trees. In *Bush base: forest farms: culture, environment and development*, E. Croll & D. Parkin (eds), 57–75. London: Routledge.

VanLiere, K. D. & R. E. Dunlap 1980. The social bases of environmental concern: a review of hypotheses, explanations and empirical evidence. *Public Opinion Quarterly* **1980**, 181–97.

Vitug, M. D. 1993. *The politics of logging: power from the forest*. Manila: Philippine Centre for Investigative Journalism.

Vogler, J. 1995. *The global commons: a regime analysis*. Chichester: John Wiley.

Vogler, J. & M. F. Imber (eds) 1996. *The environment and international relations*. London: Routledge.

Walker, B. H. 1993. Rangeland ecology: understanding and managing change. *Ambio* **22**(2–3), 80–87.

Walker, K. J. 1989. The state in environmental management: the ecological dimension. *Political Studies* **37**, 25–38.

Walker, R. B. J. & S. H. Mendlovitz (eds) 1990. *Contending sovereignties – redefining political community*. Boulder, Colorado: Lynne Rienner.

Waller, M. & F. Millard 1992. Environmental politics in Eastern Europe. *Environmental Politics* **1**(2), 159–85.

Wapner, P. 1995. Politics beyond the state: environmental activism and world civic politics. *World Politics* **47**, 311–40.

Ward, N. & P. Lowe 1994. Shifting values in agriculture: the farm family and pollution regulation. *Journal of Rural Studies* **10**(2), 173–84.

Warren, A. 1995. Changing understandings of African pastoralism and the nature of environmental paradigms. *Transactions of the Institute of British Geographers* **20**(2), 193–203.

Warren, K. J. (ed.) 1994. *Ecological feminism*. London: Routledge.

Wathern, P. (ed.) 1988. *Environmental impact assessment: theory and practice*. London: Unwin Hyman.

Watts, M. 1983a. *Silent violence: food, famine and peasantry in northern Nigeria*. Berkeley: University of California Press.

—1983b. On the poverty of theory: natural hazards research in context. In *Interpretations of calamity from the viewpoint of human ecology*, K. Hewitt (ed.), 231–62. Boston: Allen & Unwin.

—1993. Development I: power, knowledge, discursive practice. *Progress in Human Geography* **17**, 257–72.

Watts, M. & H. G. Bohle 1993. The space of vulnerability: the causal structure of hunger and famine. *Progress in Human Geography* **17**, 43–67.

WCED [World Commission on Environment and Development] 1987. *Our common future*. Oxford: Oxford University Press.

Weale, A. 1992. *The new politics of pollution*. Manchester: Manchester University Press.

Weir, D. 1988. *The Bhopal syndrome*. London: Earthscan.

Welford, R. (ed.) 1996. *Corporate environmental management – systems and strategies*. London: Earthscan.

Whitby, M. (ed.) 1994. *Incentives for countryside management: the case of environmentally sensitive areas*. Wallingford: CAB International.

—(ed.) 1996. *The European environment and CAP reform: policies and prospects for conservation*. Wallingford: CAB International.

Whitby, M. & P. Lowe 1994. The political and economic roots of environmental policy in agriculture. In *Incentives for countryside management: the case of environmentally sensitive areas*, M. Whitby (ed.), 1–24. Wallingford: CAB International.

White, L. 1967. The historical roots of our ecological crisis. *Science* **155**, 1203–1207.

White, R. R. 1994. *Urban environmental management*. Chichester: John Wiley.

Whitmore, T. C. 1990. *An introduction to tropical rain forests*. Oxford: Oxford University Press.

Wiersum, K. F. 1995. 200 years of sustainable forestry: lessons from history. *Environmental Management* **19**(3), 321–9.
Williams, M. 1989. *Americans and their forests – a historical geography*. Cambridge: Cambridge University Press.
—1990. Forests. In *The Earth as transformed by human action – global and regional changes in the biosphere over the past 300 years*, B. L. Turner, W. C. Clark, R. W. Kates, J. F. Richards, J. T. Mathews, W. B. Meyer (eds), 179–202. Cambridge: Cambridge University Press.
Wilson, G. A. 1992a. A survey on attitudes of landholders to native forest on farmland. *Journal of Environmental Management* **34**(2), 117–36.
—1992b. *The urge to clear the "bush": a study of native forest clearance on farms in the Catlins District of New Zealand, 1861–1990*. Christchurch: Canterbury University Press.
—1993. The pace of indigenous forest clearance on farms in the Catlins District, SE South Island, New Zealand, 1861–1990. *New Zealand Geographer* **49**(1), 15–25.
—1994a. German agri-environmental schemes I – a preliminary review. *Journal of Rural Studies* **10**(1), 27–45.
—1994b. Wood chipping of indigenous forest on private land in New Zealand 1969–1993. *Australian Geographical Studies* **32**(2), 256–73.
—1994c. Towards sustainable management of natural ecosystems on farms? A New Zealand perspective. *Journal of Environmental Planning and Management* **37**(2), 169–85.
—1995. German agri-environmental schemes II – the MEKA programme in Baden–Württemberg. *Journal of Rural Studies* **11**(2), 149–59.
—1996. Farmer environmental attitudes and ESA participation. *Geoforum* **27**(2), 115–31.
—1997 (in press). Factors influencing farmer participation in the environmentally sensitive areas scheme. *Journal of Environmental Management* **49**.
Wilson, O. J. 1994. "They changed the rules" – farm family responses to agricultural deregulation in South Island, New Zealand. *New Zealand Geographer* **50**(1), 3–13.
Witte, J. 1992. Deforestation in Zaire: logging and landlessness. *The Ecologist* **22**, 58–64.
Wolf, E. R. 1982. *Europe and the people without history*. Berkeley: University of California Press.
World Bank 1992. *World Development Report 1992 – development and the environment*. Washington DC: World Bank.
World Rainforest Movement & Sahabat Alam Malaysia 1989. *The battle for Sarawak's forests*. Penang: WRM and SAM.
World Resources Institute 1995. *World resources 1994–95: a guide to the global environment*. Oxford: Oxford University Press.
Worster, D. 1993. *The wealth of nature: environmental history and the ecological imagination*. Oxford: Oxford University Press.
Wright, R. V. S. 1986. New light on the extinction of the Australian megafauna. *Proceedings of the Linnean Society of NSW* **109**, 1–9.
Wynn, G. 1977. Conservation and society in late nineteenth century New Zealand. *New Zealand Journal of History* **11**(2), 124–36.
—1979. Pioneers, politicians and the conservation of forests in early New Zealand. *Journal of Historical Geography* **5**(2), 171–88.

Xu, Z., D. P. Bradley, P. J. Jakes 1995. Measuring forest ecosystem sustainability: a resource accounting approach. *Environmental Management* **19**(5), 685–92.

Yin, Y. & J. T. Pierce 1993. Integrated resource assessment and sustainable land use. *Environmental Management* **17**(3), 319–28.
Young, E. 1992. Hunter–gatherer concepts of land and its ownership in remote Australia and North America. In *Inventing places – studies in cultural geography*, K. Anderson &

F. Gale (eds), 255–72. Melbourne: Longman Cheshire.
Young, O. R. 1989. *International cooperation: building regimes for natural resources and the environment*. Ithaca, New York: Cornell University Press.
Zimmerer, K. S. 1994. Human geography and the "new ecology": the prospect and promise of integration. *Annals of the Association of American Geographers* **84**(1), 108–125.

INDEX